Lecture Notes in Economics and Mathematical Systems

444

Springer
Berlin
Heidelberg
New York
Barcelona
Budapest
Hong Kong
London
Milan
Paris
Santa Clara
Singapore
Tokyo

Steffen Jørgensen   Georges Zaccour (Eds.)

Dynamic Competitive Analysis in Marketing

Proceedings of the International Workshop
on Dynamic Competitive Analysis in Marketing,
Montréal, Canada, September 1–2, 1995

Springer

Editors

Prof. Dr. Steffen Jørgensen
Department of Management
Odense University
DK-5230 Odense M, Denmark

Prof. Dr. Georges Zaccour
GERAD, HEC
5255 Decelles Avenue
Montréal, Canada H3T 1V6
Canada

658.8
I6)2d
1995

Library of Congress Cataloging-in-Publication Data

International Workshop on Dynamic Competitive Analysis in Marketing
   (1995 : Montréal, Québec)
   Dynamic competitive analysis in marketing : proceedings of the
International Workshop on Dynamic Competitive Analysis in Marketing,
Montréal, Canada, September 1-2, 1995 / Steffen Jørgensen, Georges
Zaccour, eds.
      p.   cm. -- (Lecture notes in economics and mathematical
systems ; 444)
   Includes bibliographical references (p.).
   ISBN 3-540-61613-6 (alk. paper)
   1. Marketing--Management--Mathematical models--Congresses.
2. Competition--Mathematical models--Congresses.   I. Jørgensen,
Steffen, 1942-   . II. Zaccour, Georges. III. Title.   IV. Series.
HF5415.13.I6   1995
658.8--dc20   96-29386
                                                           CIP

ISSN 0075-8442
ISBN 3-540-61613-6 Springer-Verlag Berlin Heidelberg New York

© Springer-Verlag Berlin Heidelberg 1996
Printed in Germany

The use of general descriptive names, registered names, trademarks, etc. in this publication does not imply, even in the absence of a specific statement, that such names are exempt from the relevant protective laws and regulations and therefore free for general use.

Typesetting: Camera ready by author
SPIN: 10546888   42/3142-543210 - Printed on acid-free paper

PREFACE

This volume contains a selection of papers that were presented at the *International Workshop on Dynamic Competitive Analysis*, held in Montréal, Canada, September 1-2, 1995. The workshop was organized by the editors of the proceedings volume.

The proceedings contain both "full papers" and shorter pieces, to be considered as "work in progress". The choice of a rather broad theme for the workshop was deliberate and done in order to attract researchers from different areas of the marketing science community that usually do not get together. Obviously, a volume like this cannot be exhaustive in the coverage of the dynamics of marketing competition but we are confident that it will convey to the reader an impression of what are the current themes in this field of research. The book should be useful to researchers in marketing science, applied game theorists, graduate students, as well as practitioners in marketing with an interest in methods and examples of dynamic competitive analysis.

In Section 1, *Breton, Yezza* and *Zaccour* study dynamic advertising strategies in a differential game in a duopolistic market. The market share dynamics are the Lanchester model and the authors look for feedback Stackelberg equilibria. Due to analytical intractability, an algorithm is designed to compute equilibria with various hypothesized sets of values of the model parameters. The Lanchester model is also the subject of the study of *Erickson* who uses data from the US ready-to-eat cereal market to compare dynamic advertising strategies in closed-loop Nash equilibria with strategies based on dynamic conjectural variations. *Monahan* and *Sobel* analyze a stochastic difference game of oligopolistic competition using the notion of advertising goodwill stocks. The profits of a firm have a market share attraction form. A main purpose of the analysis is to compare advertising expenditure levels over time of risk-sensitive oligopolists with those of risk-neutral counterparts.

Section 2 starts with the paper by *Hruschka* and *Natter* who consider nonlinear models with dynamic reference prices to explain market share responses. Using a scanner data base of prices and sales of seven brands of a repeat purchase good, the authors show that their model leads to better fits as compared to models that view reference prices as expectations. *Kim* studies the question whether in a dynamic context consumers respond to price promotions in the same way for high-quality brands as for brands of lower quality. Employing data from the US market for subcompact cars, a multiple time series analysis examines the dynamic nature of such asymmetries. The paper by *Natter* and *Hruschka* studies profit impacts of agressive or cooperative pricing in a dynamic oligopoly. A reference price model is used to predict market shares. *Thépot* addresses the Coase conjecture: could a durable goods monopolist lose his monopoly because consumers expect a decrease of the price of the durable and hence are unwilling to pay more to get the product earlier? The problem is cast both as a two-period game and a differential game.

In Section 3, the paper by *Dockner, Gaunersdorfer* and *Jørgensen* is concerned with government subsidies to enhance the diffusion of a new consumer durable, say, an energy saving product. The government acts as a Stackelberg leader in a differential game in which feedback and open-loop equilibria are identified. *Minowa* and *Choi* address the problem of duopoly firms that introduce two related products: a primary product and a nondurable contingent product. The use of the former requires the latter, and diffusion and sales of the latter product are contingent upon the diffusion of the former. A Nash differential game with open-loop information is analyzed.

Channels of distribution is the subject of Section 4 in which *Bandyopadhyay* and *Divakar* study the problems of brand vs. category management. In a manufacturer-retailers channel, the assumption is that the manufacturer acts as manager of his own brand whereas retailers act as a product category manager. Using a game theoretic model, various scenarios of brand/catagory management are analyzed. Addressing the problem of determining slotting allowances, *Desiraju* develops a game from the perspective of a retailer who is dealing with a heterogeneous group of manufacturers. He identifies a sequential equilibrium when both parties have private information. The paper by *Jørgensen* and *Zaccour* discusses, although briefly, some issues that should be considered when channel members contemplate cooperation. A differential game of pricing and advertising is analyzed to identify a feedback Nash equilibrium that is to serve as the disagreement outcome in the negotiations between manufacturer and retailer.

In Section 5 various approaches to competitive analysis are presented. *Cressman* takes his starting point in the "market myopia" observed by some managers and seeks to explain why these managers act in such a way. The paper suggests a more holistic view on competition in order to identify and focus on competition on broader levels than just the strategic group to which the firm belongs. *Dearden, Lilien,* and *Yoon* take their starting point in an observation that high capital investment industries often see regular cycles of redundant capacity followed by lack of capacity. In a game theoretic model the authors show that such cyclical behavior can be rationalized as an equilibrium, even in situations where demand and output prices are stable over time. Implications for individual firms and for industrial policy are discussed. The paper by *Jaffe* provides a computer-based method for the planning of advertising and promotion expenditures in a firm, taking interdepartmental cost and efficiency aspects into account. The outcomes of a series of simulation experiments are reported. *Napiorkowski* and *Borghard* conclude the volume with a case study of the dynamics of consumer response to a marketing campaign of a long distance telephone service company, aimed at breaking the monopoly of a local telephone service provider. The study illustrates the applicability of the classification and regression tree algorithm. A model was estimated from observations of almost 300,000 customers being the targets of the campaign.

The editors would like to thank the authors for their contributions to this volume. For each paper presented at the workshop a discussant was assigned and undoubtedly the papers in this volume have benefited from the comments of the discussants as well as the plenary discussions and more informal exchanges of ideas.

As organizers of the workshop we wish to thank the following for their financial and/or logistic support of the workshop:

Centre d'Études en Administration Internationale (CETAI) of École des Hautes Études Commerciales de Montréal (HEC),

Groupe d'Études et de Recherche en Analyse des Décisions (GERAD), a joint research group of HEC, École Polytechnique de Montréal and the Faculty of Management of McGill University, Montréal,

La direction de la recherche, HEC,

The International Society of Dynamic Games,

Department of Management, Odense University, Denmark.

Finally we thank Carole Dufour, Micheline Trudeau, and Josée Vincelette for their valuable assistance with the workshop.

Steffen Jørgensen   Georges Zaccour

May, 1996

TABLE OF CONTENTS

ADVERTISING

PRICING

NEW PRODUCT DIFFUSION MODELS

FEEDBACK STACKELBERG EQUILIBRIA IN A DYNAMIC GAME OF ADVERTISING COMPETITION: A NUMERICAL ANALYSIS[1]

Michèle Breton, Abdelwahab Yezza and Georges Zaccour

GERAD, École des Hautes Études Commerciales, Montréal, Canada

Abstract

We study in this paper dynamic equilibrium advertising strategies in a duopoly with asymmetric information structure. The advertising model of Lanchester is used in a game where the relevant solution concept is feedback Stackelberg equilibrium. An algorithm is devised for the computation of this equilibrium and numerical results are reported and discussed for different values of the various parameters.

Key words: Stackelberg equilibrium, advertising, Lanchester model, dynamic game.

1   Introduction

The use of differential game models as analytical tool to investigate optimal competitive dynamic advertising decisions has become increasingly popular during the last few years. The reason is simply due to the fact that their optimal control counterparts have left behind the very important issue of competition. Further, most of earlier differential game works adopted an open-loop information structure, far less conceptually attractive than the closed-loop one which has driven most of the recent attention to this field.

---

[1]   Research supported by CETAI, HEC and NSERC-Canada.

In his survey of some differential games in advertising, Jørgensen (1982) notes that the two most important dynamic models of advertising are Lanchester's, a model based on the combat problem recognized by Kimball (1957), and Vidale-Wolfe's (Vidale and Wolfe (1957)), which are in fact quite similar (Little (1979)). Most of optimal control and differential game models proposed subsequently are more or less direct extensions of these two models.

We adopt in this paper the model of Lanchester to study optimal dynamic advertising strategies in a duopoly characterized by an asymmetric information structure, which constitutes the major distinction with respect to published literature. Indeed, we assume that the industry is formed of two firms, one of them being designated the leader and the other the follower, where, for each period, the follower sets its advertising budget after the leader has announced and committed its own. The relevant solution concept is then feedback Stackelberg equilibrium. Although this equilibrium has the very desirable property of being sub-game perfect, a drawback is that the solution cannot be characterized analytically, except in very special cases. For that reason, we devise an algorithm to compute numerical solutions, that is feedback Stackelberg equilibrium, and we run it on a number of test cases in order to assess the impact of various parameter values on equilibrium advertising strategies and payoffs.

The advertising model of Lanchester and its extensions have been adopted in numerous papers. We shall not attempt to review them and refer the interested reader to the surveys by Sethi (1977), Little (1979), Jørgensen (1982), Erickson (1991, 1995a) and Feichtinger et al. (1994). Of particular interest are those papers applying dynamic advertising models to empirical data, an avenue which is particularly needed to improve understanding of dynamic advertising decision making (e.g. Chintagunta and Vilcassim (1992), Erickson (1992, 1995b) and Fruchter (1995)).

The paper is organized as follows: Section 2 states the discrete version of the model of Lanchester in a duopoly. Section 3 is devoted to the equilibrium concept and addresses existence and computation of this equilibrium. Numerical results are discussed in Section 4 and Section 5 concludes.

2 A discrete version of Lanchester's model

In a duopoly setting, the Lanchester model assumes a competition for market share between two players. Each competitor maximizes a payoff functional, typically the total sum of her discounted profits. Denote by m_t, $0 \leq m_t \leq 1$, the market share of the leader at period t, $t=1,...,T$, where T is the planning horizon, assumed the same for both players. Let a_{it} denote the advertising level of player i, $i \in \{l(eader), f(ollower)\}$. The evolution of the leader's market share is governed by the following difference equation:

$$m_{t+1} - m_t = (1-m_t)c_l(a_{lt}) - m_t c_f(a_{ft}), \qquad (1)$$
$$m_0 \quad \text{given},$$

where we suppose that the functions c_i, $i \in \{l,f\}$ satisfy the following conditions:

$$c_i(\cdot) \geq 0; \quad \lim_{a \to \infty} c_i(a)=1; \quad c_i'(\cdot) \geq 0; \quad c_i''(\cdot) \leq 0. \qquad (2)$$

The first two conditions in (2) insure that players' market shares are bounded between 0 and 1. The last condition states that advertising is subject to decreasing marginal return.

We assume that each competitor seeks to maximize the sum of her discounted stream of profits π_i, $i \in \{l,f\}$, defined by:

$$\pi_l = \sum_{t=0}^{T} \rho_l^t \, (g_l \, m_t - a_{lt}) + b_l(m_{T+1})$$

$$\pi_f = \sum_{t=0}^{T} \rho_f^t \, (g_f \, (1-m_t) - a_{ft}) + b_f(m_{T+1})$$

(3)

where
g_i is the gross margin competitor $i \in \{l, f\}$ realizes during one time period per point of market share,

ρ_i is the discount factor for competitor $i, 0 \leq \rho_i \leq 1, \, i \in \{l, f\}$,

$b_i(\cdot)$ is the bequest function for competitor i, with $b_i(\cdot) \geq 0$ and $b_i'(\cdot) \geq 0, \, i \in \{l, f\}$.

3   Equilibrium concept and computation

The duopoly model formulated in Section 2 is a two-person non-zero sum dynamic game with two competing firms as players. The particular equilibrium concept to be considered here is based on the following assumptions concerning the way in which the players determine their advertising policies. At decision time t, the players observe the current state of the system, that is, the market shares at the beginning of period t. Player l, the leader, first chooses an advertising level $a_{lt} \geq 0$ for period t. Player f, the follower, observes the advertising level set by the leader and then chooses her own advertising level $a_{ft} \geq 0$ for period t. Each player i then receives a reward $r_i(a_{it}, m_t)$, for period t, which depends on her own advertising decision and on the state of the system m_t,

$$r_l(a_{lt}, m_t) = g_l \, m_t - a_{lt}$$

$$r_f(a_{ft}, m_t) = g_f \, (1-m_t) - a_{ft}.$$

(4)

The system then moves to a new state m_{t+1} according to transition law (1), which is controlled by the actions of both players. The new state m_{t+1} is observed by the players at the beginning of time period $t+1$, and so on until horizon T is reached. At the end of period T, the players receive each a terminal reward which depends on the final market shares. Each competitor i seeks non negative values for a_{it}, $t=1,...,T,$, maximizing her long term profit, as defined by (3), subject to (1).

This defines a sequential game with an asymmetrical information structure, where, at each stage, the follower chooses her action with complete knowledge of the leader's action. The appropriate sub-game perfect solution concept for this sequential game is the feedback Stackelberg equilibrium, where a two-level optimization is performed at each decision time, and where, at each stage of the game, no player can benefit by unilaterally deviating from its equilibrium strategy.

A feedback strategy prescribes a non-negative advertising amount to competitor i as a function of the information which is available to her at the decision time. A feedback strategy δ for competitor l is a function associating an admissible advertising amount to each state of the system for each period,

$$\delta : (m_t,t)\in[0,1]\times[0,T] \rightarrow \delta(m_t,t)\in\mathbb{R}^+. \qquad (5)$$

Let Δ be the set of feedback strategies for player l. A feedback strategy γ for competitor f is a function associating a non-negative advertising amount for each pair (m_t,a_{lt}) for each period,

$$\gamma : (m_t,a_{lt},t)\in[0,1]\times\mathbb{R}^+\times[0,T] \rightarrow \gamma(m_t,a_{lt},t)\in\mathbb{R}^+. \qquad (6)$$

Let Γ be the set of feedback strategies for player f.

Let $\pi_i(\delta,\gamma)$ denote the total discounted sum of profit over the planning horizon for player i when the players use the strategy pair (δ,γ).

Then a strategy pair (δ^*, γ^*) is a feedback Stackelberg equilibrium if

$$\pi_l(\delta^*, \gamma^*) \geq \pi_l(\delta, \gamma^*) \qquad \forall \delta \in \Delta$$

$$\pi_f(\delta^*, \gamma^*) \geq \pi_f(\delta^*, \gamma) \qquad \forall \gamma \in \Gamma.$$

The feedback Stackelberg solution was initially proposed by Simaan and Cruz (1973) for multi-stage games. Başar and Haurie (1984) showed that a feedback Stackelberg equilibrium could be related to a feedback Nash equilibrium for an associated expanded game with twice as many stages. This result has been extended to sequential games by Breton et al. (1988), who showed the existence of such solutions in a large class of zero-sum sequential games. Breton et al. (1994) proved the existence under mild conditions on the problem's parameters in a non-zero-sum pricing game of a new product with dynamic demand and cost functions.

In the context of the duopoly game considered here, the associated expanded game is played as follows. At stage $k=2t$, player l observes the state of the system m_t and chooses an advertising level a_{lt} for her product during period t; player f does not play at this stage. The system then moves to a new state (m_t, a_{lt}). At the next stage $k=2t+1$, player f observes the state of the system and chooses an advertising level a_{ft} for her product during period t; player l does not play. Each player then receives the reward $r_i(a_{it}, m_t)$ and the system moves to a new state m_{t+1} according to the transition law (2).

It is easy to show, using a dynamic programming argument, that a strategy pair defined by $[\delta^*, \gamma^*]$ is a feedback Nash equilibrium in the expanded game, corresponding to a feedback Stackelberg equilibrium in the original game, if there exist functions $\delta^*:[0,1]\times[0,T]\to\mathbb{R}^+$, $\gamma^*:[0,1]\times\mathbb{R}^+\times[0,T]\to\mathbb{R}^+$, $v_i:[0,1]\times[0,T+1]\to\mathbb{R}$ and $w_i:[0,1]\times\mathbb{R}^+\times[0,T]\to\mathbb{R}$ for $i\in\{l,f\}$ such that the following system (7)-(13) is satisfied for all $m\in[0,1]$, $y\in\mathbb{R}^+$ and $t=1,...,T$.

$$v_i(m,T+1) = b_i(m), \ i \in \{l,f\} \tag{7}$$

$$w_f(m,y,t) = \max_a \ \{r_f(a,m) + \rho_f v_f(m',t+1)\} \tag{8}$$

$$\text{s.t.} \qquad m' = m + (1-m)\,c_l(y) - m\,c_f(a)$$

$$\gamma^*(m,y,t) \in \text{arg max } \{\text{Problem (8)}\} \tag{9}$$

$$w_l(m,y,t) = r_l(y,m) + \rho_l v_l(m',t+1) \tag{10}$$

$$\text{where} \qquad m' = m + (1-m)\,c_l(y) - m\,c_f(\gamma^*(m,y,t))$$

$$v_l(m,t) = \max_a \ \{w_l(m,a,t)\} \tag{11}$$

$$\delta^*(m,t) \in \text{arg max } \{\text{Problem (11)}\} \tag{12}$$

$$v_f(m,t) = w_f(m,\delta^*(m,t),t). \tag{13}$$

Generalized Stackelberg strategy concepts have been proposed to extend the Stackelberg solution to cases where the follower has non unique rational response (see Leitman (1979)). In the same spirit, additional conditions can be added to (7)-(13) in order to define a selection of the follower's optimal strategy in (9), assuming a lexicographic ordering for the followers optimal actions, first in terms of her own profit and second in terms of the leader's profit. However, even in that case, existence of a Nash equilibrium in the expanded game does not imply existence of a Stackelberg equilibrium in the original game (Breton et al. (1988)).

In order to obtain existence results, we make the following assumption:

8

Assumption A1: The parameters of the duopoly game are such that, at each stage of the game, the solutions to problems (8) and (11) are unique, for all $m \in \{0,1\}$ and all $y \in \mathbb{R}^+$.

Proposition 1 *Under Assumption A1, there exists a feedback Stackelberg equilibrium for the duopoly model defined in Section 2.*

Proof

First notice that, since $0 \leq m_t \leq 1$, player i can always secure a non-negative sum of discounted profits π_i, for any initial m_0, by choosing $a_{it}=0$ for $t=1,...,T$. It is easy to show that π_i is bounded above by

$$\frac{1-\rho^T}{1-\rho}g_i + \rho^T b_i(1) = A_i.$$

Therefore, if, at any time period t, player i chooses an advertising level a_{it} such that $a_{it} > g_i + A_i$, then the sum of discounted profits for player i will be negative. We can thus restrict the admissible advertising levels in the optimization problems (8) and (11) to the compact sets $[0, g_i + A_i]$, $i=f, l$ respectively.

Assume that at $t+1$, v_l and v_f are continuous in m. Then, continuity of c_l, c_f and r_f imply, since the optimization in (8) is made over the compact set $[0, g_f + A_f]$, that w_f is well defined and continuous in (m,y) at t. Assumption A1 implies that, at t, γ^* is a well defined continuous function of (m,y).

Continuity of r_l, c_l, c_f, γ^* at t and v_l at $t+1$ imply continuity of w_l in (m,y) at t, which in turn imply, under Assumption A1, that the functions v_l, δ^* and v_f are well defined and continuous in m at t.

Existence of a feedback Stackelberg equilibrium follows from the continuity of the bequest functions b_i since then v_l and v_f are continuous at $T+1$. ∎

The dynamic program (7)-(13) defines a sequential Stackelberg equilibrium in the duopoly game. An approximation to this equilibrium can be found by using an approximation procedure to solve the dynamic program (7)-(13). This can be achieved by imposing a finite grid M^k on [0,1] and A_l^k on $[0, g_l + A_l]$ and defining an interpolation scheme for the computation of $v_i(m,t)$ for $(m) \notin M^k$.

One possible approach is to choose an interpolation scheme that generates continuously differentiable functions v_f and that provides estimates of derivatives $\partial v_f / \partial m$. The solution of Problem (8) at t can then be obtained by solving

$$\frac{d}{da} \{r_f(a,m) + \rho_f v_f(m'(m,y,a),t+1)\} = 0 \qquad (14)$$

for $(m,y) \in M^k \times A_l^k$. Notice however that the function $r_f + \rho\, v_f$ is generally not concave in a and that the solution of (14) may yield multiple local optima.

Another possible approach is to impose a finite grid A_f^k on $[0, g_f + A_f]$ and to solve problem (8) by simple enumeration. In that case, a cruder interpolation scheme can be used for v_l and v_f, such as one where these functions take constant values on subsets of M.

We used this second approach for the numerical illustrations presented in section 4. The interpolation scheme we propose is then the following: for m^1 and $m^2 \in M^k$, consider m' such that $m^1 < m' < m^2$. The approximate value of $v_i(m',t)$ is given by the following linear interpolation:

$$\bar{v}_i(m',t) = v_i(m^1,t) + \frac{m'-m^1}{m^2-m^1}\left(v_i(m^2,t) - v_i(m^1,t)\right).$$

4   Numerical results

In this section, we present a sample of computational results to illustrate the kind of insights that can be obtained using the model and assess the impact of main parameters' values on market share and advertising trajectories.   The following functional form is chosen for $c_i(a_{it})$, $i \in \{l, f\}$

$$c_i(a_{it}) = \beta_i \left[\frac{a_{it}}{1 + a_{it}} \right]^{\alpha_i}, \quad 0 \le \alpha_i \le 1, \quad 0 \le \beta_i \le 1$$

which satisfies (2) under the restrictions imposed on the parameters.

Further, we assume that the bequest functions are identically equal to zero.   This assumption will imply that advertising will decline to zero at the end of the horizon.   When the discount factor is strictly less than 1, considering a relatively long horizon is sufficient for the bequest function having a negligible impact on advertising strategies in the first periods.   Thus, optimal stragegies obtained by a "rolling horizon" procedure should conincide in the first periods.   Because of the contraction property of the dynamic programming operator, infinite horizon solutions can be obtained by iteratively using the functions $v_i(\cdot, 1)$ as the bequest functions $b_i(\cdot)$.

The   model   includes   10   parameters,   namely T, m_0, ρ_i, α_i, β_i, g_i, $i=l, f$.   To keep the number of figures manageable, we assume that both players discount their profits at the same market rate, taking $\rho = 0.9$.   Further, as in Erickson (1991), we assume also that both players' gross margins are equal.   Results, that is, the evolution of advertising and market shares, are presented for an horizon of 10 periods, which suffices with our choice of parameters to obtain stationarity.

We shall refer to the following values of the parameters as the base case:

$T=10$, $m_0=0.5$, $\rho_l=\rho_f=0.9$, $g_l=g_f=100$, $\alpha_i=\beta_i=0.5$, $i=l, f$.

The results of variations in parameters m_0, α_i and β_i w.r.t. the base case will be depicted and analysed.

Our investigations are identified with two sets of results. In the first one, we assume symmetry among the players (both have the same parameters' values) and we do some comparative statics, i.e. we vary one parameter at a time, all others remaining at their base case levels. In the second set, we consider some asymmetric cases, where one player has an advantage on her competitor. In both sets, we alternate the roles of the leader and the follower to assess the impact, if any, of players' labels on their profits.

4.1 Symmetric cases

In this setting, both players exhibit exactly the same parameters' values, but possibly different initial market shares.

4.1.1 Impact of initial market share

Figure (1a) shows the evolution of the market share of the leader for different values of m_0. Notice that the equilibrium market share is reached quickly (after very few periods). This result points out to the fact that initial market shares position has not durable effects. An examination of the advertising trajectories (figures (1b)-(1c)), clearly demontrates that the leader's advertising level, notably during the first periods i.e. before reaching the equilibrium level, is a decreasing function of her market share. The reverse is true for the follower i.e. her advertising level is an increasing function of leader's market share. As expected, each player's sum of discounted profits is an increasing function of her initial market share. These results are intuitively appealing. Indeed, one expects a firm to increase its advertising rate precisely when this action is badly needed, i.e. when its market share is low, and to earn a higher profit with higher market share. Finally, we observe, as expected, that advertising levels decrease when approaching the final date of the game and become zero at T. As we pointed out earlier, this is due to the finite horizon, no salvage value setting; it is also observed in all cases to be discussed below.

4.1.2 Impact of the convexity parameter

The convexity of the market share capture functions c_i is embodied in the parameter α_i. A large value of this parameter is associated to a capture function for which returns decrease more rapidly toward the asymptotic value of 1, and, as a consequence, where the capturing potency of advertising dollars is always inferior with respect to a function with a smaller α_i, significatively for relatively small advertising levels. Therefore, to attract a given additional market share, a player i needs to advertise more when α_i is high. This is observed in our tests. Indeed, figures (2b) and (2c) show that increasing the value of α lead to higher advertising levels for both players, at all periods. The impact on market share is negligible (see figure (2a)), when all parameters are equal for both players, including initial market share. As a consequence, profits decrease with α, since revenues are almost not affected whereas (advertising) cost increases.

4.1.3 Impact of effectiveness parameter

The parameter β_i is a scaling factor which represents, for player i, the effectiveness of advertising in terms of market share. Figure (3a) shows that increasing simultaneously and equally this parameter for both players has almost no impact on market shares. Interestingly, the higher is the common β, the higher are (generally) the advertising levels (figures (3b)-(3c)) and the lower are the cumulative profits: the players, in this symmetric setting, choose to be collectively less effective in their advertising strategies. This is due to the fact that higher values for the effectiveness parameter increase the competitive aspect of the game in the model of Lanchester; indeed, a higher value of β_i gives player i more leverage on the market share of the other player.

13

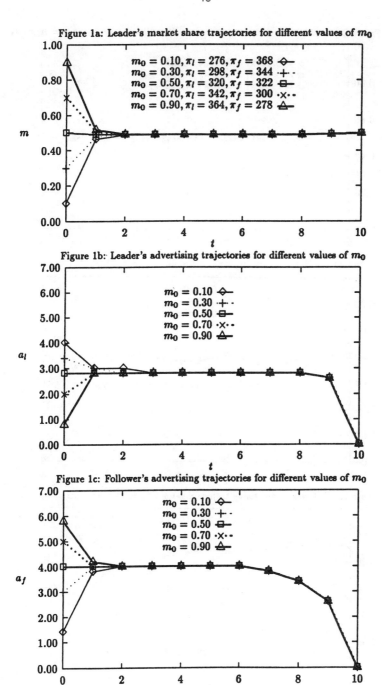

Figure 1a: Leader's market share trajectories for different values of m_0

Figure 1b: Leader's advertising trajectories for different values of m_0

Figure 1c: Follower's advertising trajectories for different values of m_0

14

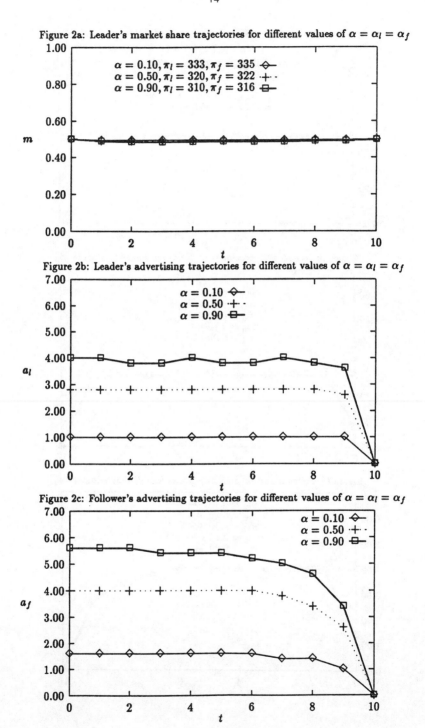

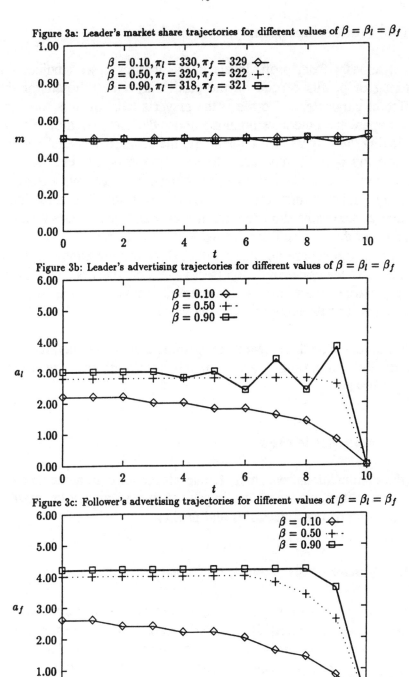

Figure 3a: Leader's market share trajectories for different values of $\beta = \beta_l = \beta_f$

$$\beta = 0.10, \pi_l = 330, \pi_f = 329$$
$$\beta = 0.50, \pi_l = 320, \pi_f = 322$$
$$\beta = 0.90, \pi_l = 318, \pi_f = 321$$

Figure 3b: Leader's advertising trajectories for different values of $\beta = \beta_l = \beta_f$

$$\beta = 0.10$$
$$\beta = 0.50$$
$$\beta = 0.90$$

Figure 3c: Follower's advertising trajectories for different values of $\beta = \beta_l = \beta_f$

$$\beta = 0.10$$
$$\beta = 0.50$$
$$\beta = 0.90$$

4.1.4 Profit comparisons

As it appears from profit levels in the various cases studied, the ranking of profits depend on the particular values chosen for the different parameters. To see if the player's label matters, one has to compare the outcomes obtained under the same conditions, e.g. by letting the players face sequentially the same parameters' values. Comparing profit levels' accordingly, it appears that the follower has a structural advantage over her competitor, securing usually a slightly higher profit than the leader in symmetric situations. Actually, the results show that the follower advertises more than the leader and thar her equilibrium market share is around 0.51. Note that profit figures are not much affected by varying simultaneously and equally the convexity and effectiveness parameters. However, initial market share has a considerable impact on the sum of discounted profits earned during the game.

To conclude on these symmetric results, one can say that they are intuitively appealing and their interpretation relatively staightforward.

4.2 Asymmetric cases

In these simulations, we shall let one player have an advantage on her competitor and assess the impact of this advantage on market shares, advertising trajectories and profits.

4.2.1 Impact of the convexity parameter

Figures (4a)-(4d) exhibit advertising trajectories for different values of α_l and α_f. Corresponding leader's market share trajectories are given in figures (4e)-(4f). From these figures, one concludes that varying convexity parameter α_i for player i does not affect much, at least during the first periods, the advertising level of her rival, but has an important direct effect on her own advertising level. The impact of varying the convexity parameter on the market share is as expected: increasing α_i for player i while keeping her competitor's parameter constant decreases player i's market share. As a consequence, the effect of increasing α_i for player i while keeping her competitor's parameter constant is a decrease of her profit and an less important increase on her competitor's profit.

4.2.2 Impact of the effectiveness parameter

Figures (5a)-(5d) exhibit the advertising trajectories for different values of β_l and β_f. The leader's market share trajectories for these values are depicted in figures (5e)-(5f). From these, we observe the following:

(i)   Varying the effectiveness parameter of one player affects the advertising levels of both players.

(ii)   The advantadged player gets a higher market share and a higher profit.

(iii)   Increasing the value of the effectiveness parameter for one player from 0.1 to 0.5 while keeping the rival's one at the base case value of 0.5 leads to higher advertising levels for both players.

(iv)   Increasing the value of the effectiveness parameter for one player from 0.5 to 0.9 while keeping the rival's one at the base case value of 0.5 leads generally to lower advertising levels.

The first observation is explained by the fact that the effectiveness parameter enters the model linearly and has a higher impact on the changes in market share of the rival player than the convexity parameter. The second observation is intuitive; the player enjoying higher effective advertising gets a higher market share and a higher profit. The last two observations seem to indicate that one player's advertising increases (*resp. does not increase and may even decrease*) with her own effectiveness parameter if its value does not exceed (*resp. exceeds*) the value of the effectiveness parameter for the rival player. This result is fairly intuitive; indeed, it means that optimal advertising levels depend on both firms effectiveness parameters (and of course on market share).

4.2.3 Profit comparisons

The same observations made in the symmetric case hold in the asymmetric context.

5 Conclusion

Falling short from being able to characterize analytically the advertising trajectories of the firms in a leader-follower dynamic game of advertising competition with feedback information stucture, the numerical approach used in this paper allows us to obtain some interesting economic insight for the problem of optimal advertising.

The next step is to assume a non-zero bequest functions for the players. Some preliminary simulations show that in the case where the salvage value is relatively important, one needs to be careful when chosing the parameter ranges. Indeed, notably for high values of one of the effectiveness parameters, it turns out that the model prescribes for the advantaged player a policy that leads to a very low market share at T-1, which is counterintuitive. Mathematically, the explanation is obvious, since that by construction the Lanchester model assumes that advertising is oriented only toward the rival's customers. Therefore, a wise player should set an advertising level that insures her a very low market share at T-1 and thus renders her able to capture a huge market share at T and thus a huge salvage reward.

19

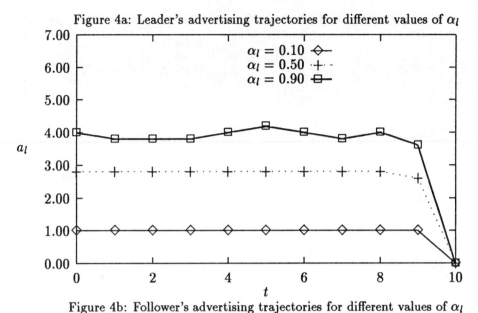

Figure 4a: Leader's advertising trajectories for different values of α_l

Figure 4b: Follower's advertising trajectories for different values of α_l

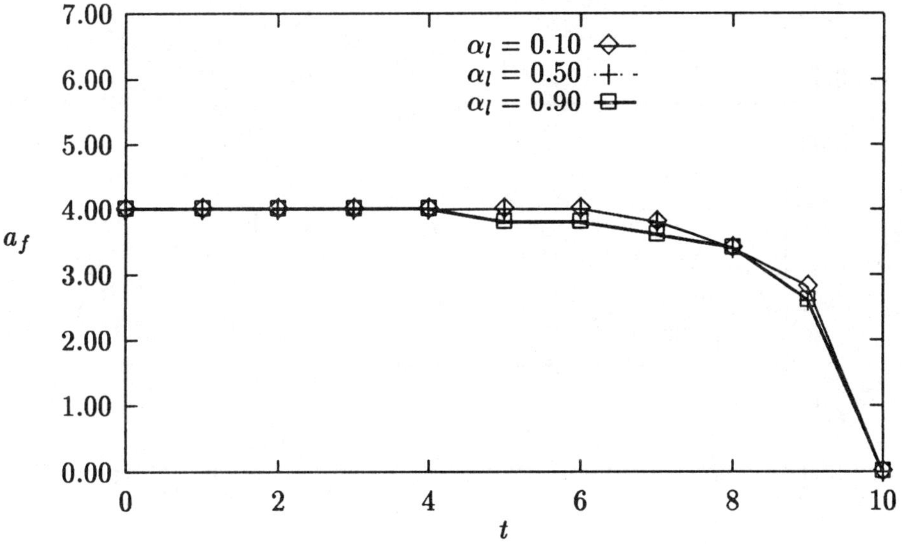

20

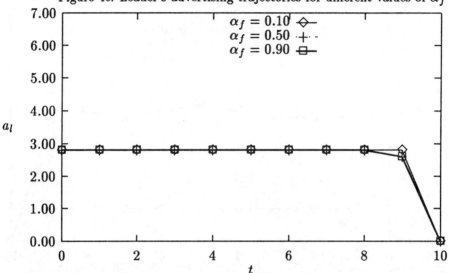

Figure 4c: Leader's advertising trajectories for different values of α_f

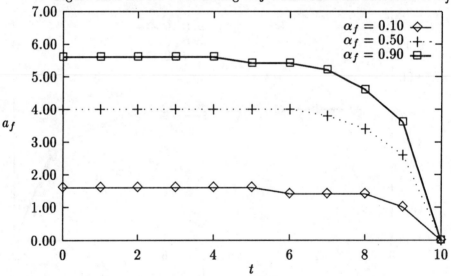

Figure 4d: Follower's advertising trajectories for different values of α_f

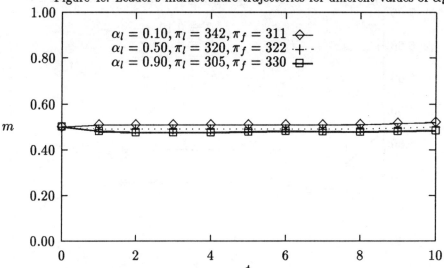

Figure 4e: Leader's market share trajectories for different values of α_l

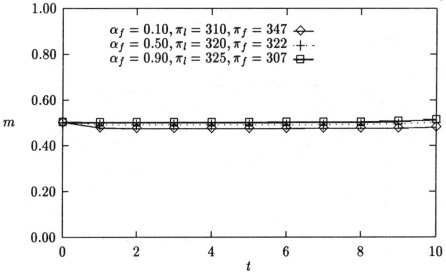

Figure 4f: Leader's market share trajectories for different values of α_f

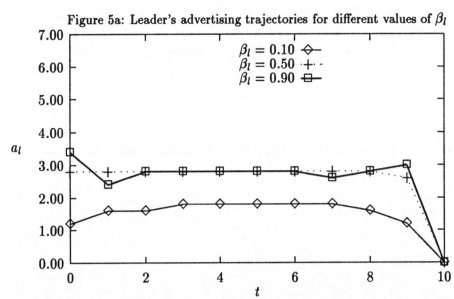

Figure 5a: Leader's advertising trajectories for different values of β_l

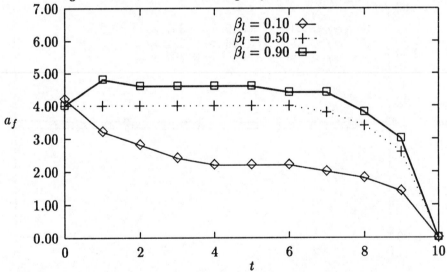

Figure 5b: Follower's advertising trajectories for different values of β_l

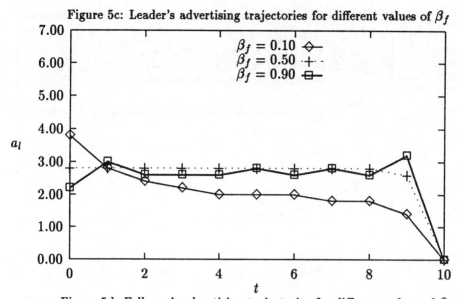

Figure 5c: Leader's advertising trajectories for different values of β_f

Figure 5d: Follower's advertising trajectories for different values of β_f

24

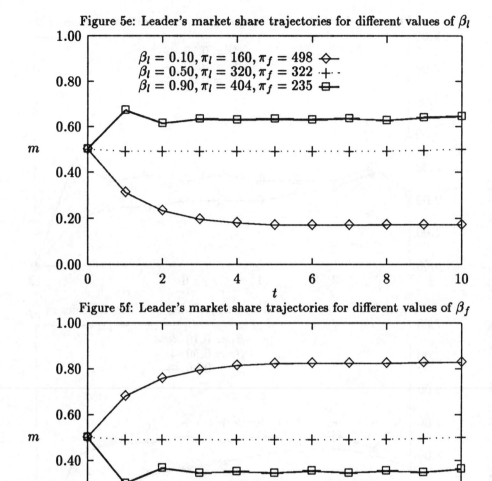

Figure 5e: Leader's market share trajectories for different values of β_l

Figure 5f: Leader's market share trajectories for different values of β_f

References

BAŞAR, T. and A. HAURIE (1984), "Feedback Equilibria in Differential Games with Structural and Modal Uncertainties", *Advances in Large Scale Systems*, Edited by J. B. Cruz, JAI Press, Greenwich, Connecticut.

BRETON, M., A. ALJ and A. HAURIE (1988), "Sequential Stackelberg Equilibria in Two-Person Games", *Journal of Optimization Theory and Applications*, **59** (October), 71-97.

BRETON, M., F. CHAUNY and G. ZACCOUR (1994), "A Leader-Follower Dynamic Game of New Product Diffusion", to appear in *Journal of Optimization Theory and Applications*.

CHINTAGUNTA, P.K. and N. VILCASSIM (1992), "An Empirical Investigation of Advertising Strategies in a Dynamic Duopoly", *Management Science*, **38** (September), 1230-1244.

ERICKSON, G.M. (1991), "Dynamic Models of Advertising Competition", *International Series in Quantitative Marketing*, Kluwer Academic Publishers.

ERICKSON, G.M. (1992), "Empirical Analysis of Closed-Loop Duopoly Advertising Strategies", *Management Science*, **38** (December), 1732-1749.

ERICKSON, G.M. (1995a), "Differential Game Models of Advertising Competition", *European Journal of Operational Research*, **83**, 431-438.

ERICKSON, G.M. (1995b), "An Empirical Comparison of Oligopolistic Advertising Strategies", mimeo.

FEICHTINGER, G., R.F. HARTL and S. SETHI (1994), "Dynamic Optimal Control Models in Advertising: Recent Developments", *Management Science*, **40** (February), 195-226.

FRUCHTER, G.E. (1995), "Feedback Promotional Strategies in a Growing Dynamic Competitive Market: Theory and Empirical Study", mimeo.

JØRGENSEN, S. (1982), "A Survey of Some Differential Games in Advertising", *Journal of Economic Dynamics and Control*, **4**, 341-369.

KIMBALL, G.E. (1957), "Some Industrial Applications of Military Operations Research Methods", *Operations Research*, **5** (April), 201-204.

LEITMAN, G. (1979), " On Generalized Stackelberg Strategies", *Journal of Optimization Theory and Applications*, **26**, 637-643.

LITTLE, J.D.C. (1979), "Aggregate Advertising Models: The State of the Art", *Operations Research*, **27** (July-August), 629-667.

SETHI, S.P. (1977), "Dynamic Optimal Control Models in Advertising: A Survey", *SIAM Review*, **19** (October), 685-725.

SIMANN, M. and J.B. CRUZ (1973), "On the Stackelberg Strategy in Nonzero-Sum Games", *Journal of Optimization Theory and Applications*, **11**, 533-555.

VIDALE, M.L. and H.B. Wolfe (1957), "An Operations Research Study of Sales Response to Advertising", *Operations Research*, **5** (June), 370-381.

An Empirical Comparison of
Oligopolistic Advertising Strategies

Gary M. Erickson, University of Washington, USA

Abstract. Two recent approaches have been proposed for developing advertising strategies in dynamic oligopolies, in the face of mathematical difficulties involved in developing closed-loop strategies in differential games. The proposed approaches, along with two base-line approaches, are compared in the context of a Lanchester oligopoly model and the empirical setting of the U. S. ready-to-eat cereal market. Three conclusions arise from the empirical comparison: (1) dynamic advertising strategies that are developed from a differential game context more closely match the general level of actual advertising expenditures than do strategies based on single-period decisions, (2) strategies based on dynamic conjectural variations provide the best overall fit with competitors' actual advertising, (3) the strategies considered do not fully capture the dynamic advertising behavior of oligopolistic competitors.

1 Introduction

The use of dynamic analysis has led to significant contributions in recent years to the study of marketing competition, particularly in the area of advertising. As a recent survey (Erickson 1995a) indicates, differential game modeling of advertising competition has allowed valuable insights to be gained through numerical, analytical, qualitative, and empirical analysis of competitive advertising strategies.

The study of dynamic advertising competition has run into difficulties, however. Much of the research to date has involved duopolistic competition, and although much has been learned about that restricted competitive situation, very little research involving dynamic analysis has been devoted to competition involving more than two competitors. A major difficulty in moving beyond duopolies to the study of more general oligopolies is a mathematical one; derivation of important closed-loop strategies, so called because such strategies allow advertising levels to vary with state variables such as market shares, becomes difficult, if not impossible, when the competitive situation involves more than one state variable, as it does when there is oligopolistic competition with three or more competitors (Starr and Ho 1969, Case 1979, Clemhout and Wan 1979).

In the face of this mathematical obstacle, two alternative modeling approaches have been suggested in the attempt to extend dynamic analysis to oligopolistic advertising competition (Erickson 1995b, c). The approaches are quite different in nature: one (Erickson 1995b) assumes single-period decision making but with salvage values to incorporate returns beyond the current period, and the other (Erickson 1995c) extends differential game methodology to involve conjectures about rival advertising reactions to changes in market share state variables. Each approach maintains elements of dynamic and closed-loop strategies, but also involves compromises in its mathematical modeling of oligopolistic strategies. As such, it is not immediately clear how useful each is in capturing the advertising patterns of oligopolistic competitors.

The present paper takes the opportunity to provide an empirical comparison of the two suggested modeling approaches, to each other, to base-line models, and to the actual advertising of oligopolistic competitors. The intent is to gain insight into the extent to which the suggested approaches can explain the advertising behavior of competitors in a dynamic oligopoly. The plan of the paper is as follows. The next section contains a description of the alternative modeling approaches advanced by Erickson (1995b, c) as well as two base-line approaches, in the context of a Lanchester model of advertising competition. This is followed by an empirical comparison of advertising strategies developed from the alternative models, in a particular oligopolistic setting involving the ready-to-eat cereal industry. Discussion follows, and a conclusions section ends the paper.

2 Model and Strategy Alternatives

A Lanchester model of market share dynamics in an oligopoly is used as the foundation for analysis and comparison of alternative advertising strategies:

$$\dot{M}_i = \beta_i \sqrt{A_i} - M_i \sum_{j=1}^{n} \beta_j \sqrt{A_j} \, , \ i = 1,\ldots,n. \tag{1}$$

where

M_i = competitor i's market share, $i = 1,\ldots,n$
A_i = competitor i's advertising rate, $i = 1,\ldots,n$.

The square-root formulation in (1) is to capture in the model diminishing marginal advertising effects. It is also necessary to work with a discrete-time version of the Lanchester model:

$$M_{it} - M_{i,t-1} = \beta_i \sqrt{A_{it}} - M_{i,t-1} \sum_{j=1}^{n} \beta_j \sqrt{A_{jt}} \, , \ i = 1,\ldots,n \tag{2}$$

where t indexes discrete time periods.

Erickson (1995b) suggests that a way to view decision-making in a dynamic market environment is to consider that competitors make single-period decisions, but that each period's decision by a competitor involves a salvage value as well as short-term profit:[1]

$$\max v_{it} = g_t M_{it} - A_{it} + w_i M_{it}$$

$$= (g_t + w_i) M_{it} - A_{it},$$

$$i = 1,\ldots,n. \tag{3}$$

In (3), g_t is competitor i's gross profit rate, and market share M_{it} is assumed to follow the discrete Lanchester pattern in (2). The salvage value coefficient w_i allows the competitor to consider the longer-term value of market share achieved in the current period, value that extends beyond the current period. In this way, an important aspect of decision making in dynamic markets is achieved, in that the decision maker considers not only the short-term but also the long-term gain from the advertising decision being made.

A Nash equilibrium is derived by setting $\partial v_{it}/\partial A_{it} = 0$ for each i while recognizing the Lanchester structure (2), and yields *salvage-value* advertising strategies:

$$A_{it}^{sv} = \frac{(g_i + w_i)^2 \beta_i^2 (1 - M_{i,t-1})^2}{4}, \ i = 1,\ldots,n. \tag{4}$$

Note that, since the advertising decision of each competitor depends on its market share lagged one period, the advertising strategies in (4) are "closed-loop-like" in the sense that each competitor's advertising level depends on its market share (lagged one period, given the discrete-time nature of the model).

An alternative, base-line, strategy is one based solely on short-term decision making, i.e. that each competitor seeks to maximize short-term profit, and ignores long-term considerations. Competitors' *short-term* advertising strategies derive from (4) by assuming $w_i = 0$, $i = 1,\ldots,n$:

$$A_{it}^{st} = \frac{g_i^2 \beta_i^2 (1 - M_{i,t-1})^2}{4}, \ i = 1,\ldots,n. \tag{5}$$

The salvage-value and short-term strategies defined in (4) and (5) are essentially static approaches, although the salvage value approach allows for returns that are anticipated to extend beyond the short term. Dynamic strategies can be developed through a differential game framework, although oligopoly analysis in a differential game setting is hampered by the mathematical difficulties associated with developing closed-loop Nash equilibria. Because of these difficulties, Erickson (1995c) suggests an approach that allows numerical computation of Nash equilibrium strategies. The approach works with the necessary conditions for Nash equilibria and involves an interpretation that considers competitors' conjectures about rival responses to changes in market shares.

A differential game interpretation involves continuous-time dynamics. In the present context, the continuous-time Lanchester model (1) is used to model the

---

[1]The model (3) is a simpler, less general, version of the model in Erickson (1995b), which has a quadratic formulation for the salvage value.

dynamics of market share evolution. Further, it is assumed that each competitor wishes to maximize discounted profits over an infinite time horizon:

$$\max \int_0^\infty e^{-rt} (g_i M_i - A_i) dt, \ i = 1,\dots,n \tag{6}$$

where r is a discount rate. Also, the advertising rate is time-variant, although for notational convenience the time subscript notation is dropped for the continuous-time model.

The Hamiltonians for the differential game are:

$$H_i = g_i M_i - A_i + \sum_{j=1}^n \lambda_{ij} (\beta_j \sqrt{A_j} - M_j \sum_{k=1}^n \beta_k \sqrt{A_k}), \ i = 1,\dots,n \tag{7}$$

Necessary conditions for a Nash equilibrium involve the costate variables λ_{ij} in the following way (see, e.g., Kamien and Schwartz 1991, Section 23):

$$\dot{\lambda}_{ij} = r\lambda_{ij} - \frac{\partial H_i}{\partial M_j} - \sum_{\substack{k=1 \\ k \neq i}}^n \frac{\partial H_i}{\partial A_k} \frac{\partial A_k}{\partial M_j}, \ i, j = 1,\dots,n. \tag{8}$$

The summed term of cross-effects on the right-hand-side of (8) is needed, since it is presumed that advertising amounts under such strategies depend on the market share state variables; the summed term is also the source of the mathematical difficulty in deriving closed-loop strategies (Fershtman 1987). To alleviate such difficulty, Erickson (1995c) proposes that the advertising relationships with market shares be interpreted as dynamic conjectural variations

$$C_{ijk} \equiv \frac{\partial A_k}{\partial M_j}, \ j,k = 1,\dots,n, \ k \neq i. \tag{9}$$

Each competitor has conjectures about how its rivals react to changes in market shares. *Dynamic-conjectural-variations* advertising strategies are derived by solving (1), (8) as a two-point boundary value problem (TPBVP),[2] with the additional necessary conditions that arise from setting $\partial H_i / \partial A_i = 0, \ i = 1,\dots,n$:

$$A_i^{dcv} = \frac{\beta_i^2 (\lambda_{ii} - \sum_{j=1}^n \lambda_{ij} M_j)^2}{4}, \ i = 1,\dots,n. \tag{10}$$

As opposed to closed-loop strategies, open-loop Nash equilibrium advertising strategies are readily computable, and can provide a base line for comparison purposes. Open-loop strategies do not consider the possibility of competitor advertising reactions to market share developments, and are the equivalent of the strategies that derive from the solution of (1), (8), and (10), but where it is assumed that the dynamic conjectural variations are all equal to zero. Specifically, *open-loop* strategies are derived through

---

[2]The beginning values of the market share variables are assumed to be known. Also, the ending values of the costate variables, at a chosen "large" value of the time period t, are set to zero.

$$A_i^{ol} = \frac{\beta_i^2(\lambda_{ii}^* - \sum_{j=1}^{n} \lambda_{ij}^* M_j)^2}{4}, \ i = 1,\ldots,n \tag{11}$$

where M_i, $i = 1,\ldots,n$, are subject to (1) and

$$\dot{\lambda}_{ij}^* = r\lambda_{ij}^* - \frac{\partial H_i}{\partial M_j}, \ i,j = 1,\ldots,n. \tag{12}$$

3 Empirical Comparison

The four types of advertising strategies defined in the previous section--short-term, salvage-value, open-loop, and dynamic-conjectural-variations--are compared empirically in the context of the ready-to-eat cereal industry. Through 1991, six cereal manufacturers competed in the industry: Kellogg, General Mills, the Post division of Philip Morris' General Foods, Quaker Oats, Ralston Purina, and RJR Nabisco. (In 1992, Philip Morris acquired the Nabisco cereal brands.) Market share and advertising data for the years 1967 through 1991 are available in various issues of *Advertising Age*; additional advertising data are obtained from various issues of *Leading National Advertisers*. Prior to analysis, the advertising data are converted to 1966 dollars.

Parameter values are needed to construct the alternative advertising strategies for comparison. Multivariate regression of the discrete version of the Lanchester model (2) produces the estimates for the advertising effectiveness parameters, β_i, $i = 1,\ldots,6$, shown in Table 1.

Table 1. Lanchester Model Parameter Estimates

Competitor	Estimate	t-Statistic
Kellogg	.00934	3.60
General Mills	.00757	4.05
Post	.00467	3.33
Quaker Oats	.00481	3.50
Ralston	.00453	4.00
Nabisco	.00391	3.38

Values for gross profit rates g_i, $i = 1,\ldots,6$, are generated by accessing financial information, specifically gross profit margins (cost of goods sold divided by sales) from annual reports of the parent companies of the cereal competitors, covering the period studied. Average gross profit margins over the period are then multiplied by the average industry sales level over the period studied to obtain gross profit rates:

$g_1 = 456$

$g_2 = 460$

$g_3 = 430$

$g_4 = 403$

$g_5 = 302$

$g_6 = 400$.

Additional parameter values are needed for certain strategies: the discount rate, salvage values, and dynamic conjectural variations. An effort is made not to favor those strategy approaches that involved additional parameters, say by determining parameter values that provide best fits with the advertising patterns. A value of .1 is assumed for the discount rate. For salvage values, it is assumed that $w_i = g_i$, $i = 1,...,6$, i.e. that each competitor anticipates it can benefit from its obtained market share one additional year. Finally, for dynamic conjectural variations, a simplifying assumption is made, that a competitor conjectures that each of its rivals reacts only to changes in its (the rival's) market share. Values for the dynamic conjectural variations are obtained from ratios of actual changes in advertising to changes in market share levels over the data period, specifically by calculating for each competitor the change in its advertising over the period divided by the change in its market share over the same period, and assigning this ratio as a dynamic conjectural variation for each of the other competitors. A final step, due to convergence problems in the computation of dynamic-conjectural-variations strategies, reduces the initial values by one-half, which yields the following values for dynamic conjectural variations:

$C_{i11} = -341$, $i = 2,...,6$

$C_{i22} = 328$, $i = 1,3,...,6$

$C_{i33} = -171$, $i = 1,2,4,5,6$

$C_{i44} = 281$, $i = 1,2,3,5,6$

$C_{i55} = 48$, $i = 1,...,5,6$

$C_{i66} = -67$, $i = 1,...,5$.

With the parameter values as determined above, a time series of values for each advertising strategy approach and for each competitor is obtained, using (4), (5), (10), and (11). The time series for the various strategies are shown in the Figure along with the actual advertising amounts. Also, sums of squared deviations fits between the strategy time series values and the actual advertising amounts are shown in Table 2.

Figure. Advertising Strategies and Actual Advertising

Kellogg

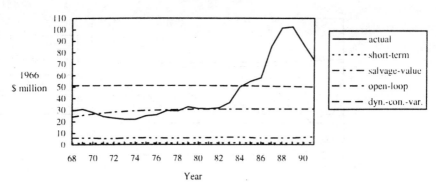

General Mills

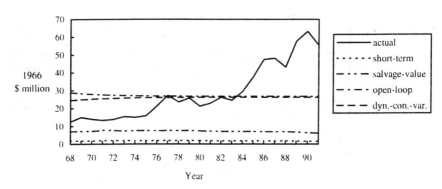

Post

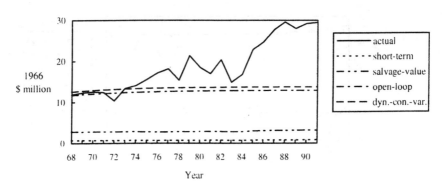

34

Figure. (Continued)

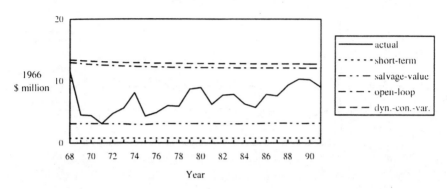

Quaker Oats

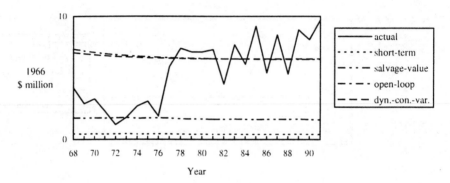

Ralston

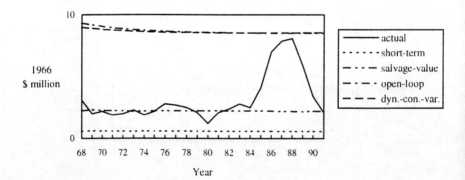

Nabisco

Table 2. Sums of Squared Deviations

Competitor	Short-Term	Salvage-Value	Open-Loop	Dynamic-Conjectural-Variations
Kellogg	59,895	51,120	19,978	17,111
General Mills	23,145	17,093	5,928	5,624
Post	8,837	7,026	1,792	1,548
Quaker Oats	1,050	479	780	944
Ralston	775	511	214	202
Nabisco	257	112	816	794
Total	93,959	76,341	29,508	26,223

A conclusion from the Figure and Table 2 is that for a majority of the cereal competitors dynamic strategies developed from the differential game structure capture the overall level of advertising expenditures more accurately than do single-period strategies. It appears that competitive advertising levels in the cereal industry are more consistent with a dynamic, long-run, approach to strategy development than with a static approach. It is true that the salvage-values strategies would fit better if larger values were permitted for the salvage values. This would make it appear that competitors consider the value from achieved market share to reach across multiple years, and reinforces the importance of building long-run considerations into advertising strategies.

Between the two dynamic strategy approaches, dynamic-conjectural-variations strategies provide a better fit to the data than do open-loop strategies. In the present analysis, a rather crude and simplistic approach is used to assign values to the dynamic conjectural variation values. There is no theoretical reason for conjectures to be nonzero only for a rival reacting to its own market share, or for the conjectures to be the same across competitors in the manner that they are assigned. A less restrictive approach to determining dynamic conjectural variations would provide even more explanatory power to advertising strategies based upon such conjectures.

Finally, it is evident from the Figure that the strategies studied show less period-to-period change than does actual advertising. The dynamic strategies generally do a decent job of capturing the overall level of advertising spending, but are unable to explain fluctuations, and in certain cases trends, in advertising expenditures. The advertising strategy approaches considered in the present study only go so far in explaining competitive advertising behavior.

4 Conclusions

The present study involves an empirical examination of two recently advanced approaches, and two base-line approaches, for developing and studying oligopolistic advertising strategies in dynamic markets. Empirical comparison indicates that competitors in the ready-to-eat cereal market appear to take dynamic considerations into account in their advertising decisions; their advertising expenditures would be at lower levels if they were being myopic. The approach based on dynamic conjectural variations provides the best overall fit with the competitors' actual advertising. The empirical results reinforce the need to study oligopolistic advertising strategies in a dynamic context, with dynamic methodologies. On the other hand, the dynamic advertising strategies studied are not effective at explaining fluctuations and trends observed in the competitors' advertising data. This presents an ongoing challenge for research involving dynamic advertising competition in oligopolies.

References

Case, James H. (1979), *Economics and the Competitive Process*, New York: New York University Press.

Clemhout, S. and H. Y. Wan, Jr. (1979), "Interactive Economic Dynamics and Differential Games," *Journal of Optimization Theory and Applications*, 27 (January), 7-30.

Erickson, Gary M. (1995a), "Differential Game Models of Advertising Competition," *European Journal of Operational Research*, 83, 431-438.

_____ (1995b), "Advertising Strategies in a Dynamic Oligopoly," *Journal of Marketing Research*, 32 (May), 233-237.

_____ (1995c), "Dynamic Conjectural Variations in a Lanchester Oligopoly," working paper.

Fershtman, Chaim (1987), "Alternative Approaches to Dynamic Games," *Global Macroeconomics: Policy Conflict and Cooperation*, Ralph C. Bryant and Richard Portes (Eds.), New York: Macmillan Press, 43-57.

Kamien, Morton I. and Nancy L. Schwartz (1991), *Dynamic Optimization: The Calculus of Variations and Optimal Control in Economics and Management* (Second Edition), Amsterdam: North-Holland.

Starr, A. W. and Y. C. Ho (1969), "Nonzero-Sum Differential Games," *Journal of Optimization Theory and Applications*, 3 (3), 184-206.

Risk-Sensitive Dynamic Market Share Attraction Games: An Extended Abstract

George E. Monahan[1] and Matthew J. Sobel[2]

[1] University of Illinois, USA
[2] SUNY at Stony Brook, USA

Abstract. We present partial results showing that risk-sensitive oligopolists would spend less on advertising than would their risk-neutral counterparts. The model is an infinite-horizon stochastic game in which, at the beginning of each time period, the level of each firm's "goodwill" is a random function of its own current and past advertising expenditures, as well as the current and past advertising expenditures of each of its competitors. The profit generated by each firm depends on its own level of goodwill and current advertising and on the goodwill levels and current advertising of each of its competitors. We assume that single-period firm profits have a market share attraction form. The objective of each firm is to maximize the expected utility of the sum of discounted profits generated over an infinite horizon. The utility function we employ allows us to model explicitly the risk sensitivity of the streams of random rewards accruing to each firm. We analyze the impact that risk sensitivity and other parameters have on equilibrium advertising strategies by exploiting the special structure of the stochastic game model.

Keywords. Risk-sensitive, stochastic game, advertising strategies

1 Introduction

Suppose that $J \geq 2$ firms compete over time on the basis of advertising in a market with a fixed sales potential of V units. The share of the market captured by each firm depends on that firm's own current and past advertising expenditures, as well as on the current and past advertising levels of each of its competitors. We assume that the share of the market accruing to each firm has an attraction form, that is, each firm's market share is the ratio of a function its own current and past advertising to the sum of a function of the current and past advertising of all of its competitors. We use the term "goodwill" to label the cumulative effect of current and past advertising. The payoff to each firm in each period is a function of the goodwill levels of each the J firms in market. We refer to the problem wherein each firm determines advertising expenditures so as to maximize expected total discounted profits in each period of an infinitely-long planning horizon as a *market share attraction game*.

Monahan and Sobel (1994) (hereafter referred to as M-S), develop and analyze a market share attraction game, wherein goodwill level of each firm in each period is a random function of the goodwill levels of all firms in the previous period, as well as the advertising expenditures of all firms in both the current and previous periods. In that model, each firm seeks to maximize the expected value of the sum of discounted profits generated over an infinite planning horizon. In the paper on which this abstract is based, we generalize that market share attraction game to consider explicitly risk-sensitive decision makers. The objective there is to determine advertising levels in each period that maximize the expected value of the *utility* of the sum of discounted profits generated over an infinite planning horizon. For each firm, we employ an exponential utility function with a parameter that is a measure of that firm's sensitivity to risk. One objective of that paper is to determine how equilibrium allocations made in this risk-sensitive environment differ from those made in the risk-neutral environment of M-S.

There is a substantial literature devoted to dynamic models of advertising competition. Virtually all of the models that have been developed in this area are differential games that focus on competition in a duopoly. Jørengen (1982) surveys the early applications of differential game theory to advertising. More recent contributions include Sorger (1989) and Erickson (1992, 1995). There is as well a large literature on the effects of risk-sensitivity but most of its content is free of any context. Its effects on structured models are investigated in relatively few studies including the recent work by Eeckhould, Gollier, and Schlesinger (1995) and its references, and in some citations near the end of Section 2.

The dynamic game model is described in Section 2 and embedded in a risk-sensitive criterion in Section 3. We show in Section 4 that the dynamic model has a *myopic* structure; that is, it corresponds to a sequence of static (one-period) games. Section 5 investigates the special cases of symmetric games and perfect competition. *Heightened risk-sensitivity in these cases leads to lower goodwill levels and advertising expenditures.*

2 The Model

Suppose that J firms compete in a fixed market over an infinite number of periods t, starting with $t = 1$. Let a_{jt} be firm j's advertising expenditure in period t and let $\mathbf{a}_t$ be the column vector $(a_{1t}, \ldots, a_{Jt})'$, where prime denotes transpose.

"Goodwill" is an aggregate measure of current and past advertising. Let g_{jt} be firm j's goodwill in period t *after* advertising is allocated in period t and let $\mathbf{g}_t = (g_{1t}, \ldots, g_{Jt})$. The model for the dynamics of goodwill is

$$\mathbf{g}_t = \mathbf{K}_{t-1}\mathbf{g}_{t-1} + \mathbf{L}\mathbf{a}_t, \tag{1}$$

where $\mathbf{K}_t$ is a random $J \times J$ matrix. The vector $\mathbf{K}_0 \mathbf{g}_0$ specifies the starting levels of goodwill and is initial data for the problem. This model stipulates that goodwill levels after advertising decisions are made in period t depend stochastically on goodwill levels at the end of period $t-1$ and the advertising done in the market in period t. Less general specifications are prevalent in the literature; see M-S for additional details and references.

The $J \times J$ matrix $\mathbf{L}$ specifies the direct impact on goodwill due to expenditures on advertising by the competitors.

It can be shown that (1) is a generalization of the goodwill dynamics specified in a single-firm model of advertising by Nerlove and Arrow (1962).

Let δ_j be firm j's single-period discount factor ($0 < \delta_j < 1$). We assume that $\mathbf{L}$ is a nonsingular matrix and that $\mathbf{M} \equiv \mathbf{L}^{-1}$ has nonnegative elements. Let ℓ_j and θ_{jt} be the j-th diagonal elements of $\mathbf{L}$ and $\mathbf{K}_{jt}$, respectively. As in M-S, we assume that $\mathbf{K}_1, \mathbf{K}_2, \ldots$ are independent and identically distributed, but $\theta_{1t}, \ldots, \theta_{Jt}$ may be correlated for fixed t.

When $\mathbf{L}$ and $\mathbf{K}_t$ are diagonal matrices, (1) simplifies to

$$g_{jt} = \theta_{j,t-1} g_{j,t-1} + \ell_j a_{jt}. \tag{2}$$

Let $\mu_j(\mathbf{g})$ be firm j's gross profit in a single period. The market share attraction form specifies that

$$\mu_j(\mathbf{g}) = \frac{V M_j b_j g_j^\beta}{\sum_{k=1}^J b_k g_k^\beta}, \tag{3}$$

where M_j is firm j's profit per unit sold and $j = 1, \ldots, J$. We assume that all parameters are positive numbers and that $\beta < 1$.

As in M-S, let $X_{jt} = \mu_j(\mathbf{g}_t) - a_{jt}$ denote firm j's net profit in period t.

The present value of firm j's profits generated over an infinite planning horizon is

$$\pi_j = \sum_{t=1}^\infty (\delta_j)^{t-1} X_{jt} = \sum_{t=1}^\infty (\delta_j)^{t-1} [\mu_j(\mathbf{g}_t) - a_{jt}]. \tag{4}$$

We assume that each firm selects its advertising levels via a nonanticipative contingency plan, called a *strategy*. The contingencies are the outcomes of stochastic elements of the model and past advertising levels of all J firms. Thus, "open-loop" decision rules and reaction functions are proper subsets of the set of all strategies.

Let $U(\pi_j)$ and $E[U(\pi_j)]$ denote the utility and the expected utility, respectively, of the discounted stream of profits π_j. In the paper, we assume that the utility function has the following form:

$$U(x) = -\exp(-\lambda x), \tag{5}$$

for any real number x. The exponential form of the utility function insures that each firm is risk-averse. In this specification, $-U''(x)/U'(x) = \lambda$, so that the parameter $\lambda > 0$, is the Pratt-Arrow measure of risk aversion that describes each firm's attitude towards risk. Exponential functions emerge naturally from fundamental utility theory considerations [Bell (1995)] and have been applied in stochastic optimization [Howard and Matheson (1972), Eagle (1975), Porteus (1975), Jaquette (1976), Denardo and Rothblum (1979), Whittle (1981), Chung and Sobel (1987), and Bouakiz and Sobel (1992)] and stochastic games [Shinde and Sobel (1991)]. A Taylor series expansion of (5) leads to the conclusion that the exponential utility function tends to risk-neutrality (a linear utility function) as λ tends to zero. It is apparent that maximizing $E[U(\cdot)]$ corresponds to minimizing $E[\exp(-\lambda\cdot)]$ and we shall henceforth employ the latter representation.

Let $V_j(d_1,\ldots,d_J\,|\,h)$ denote $E[U(\pi_j)]$ when $d_1,\ldots,d_J$ are the strategies employed by firms $1,\ldots,J$, respectively, and h specifies the initial conditions of the game (a_0, g_0, and K_0). Strategies are nonanticipative functions of elapsed histories that stipulate nonnegative advertising expenditures, i.e., $a_{jt} \geq 0$ for all firms j and time periods t. We employ a Nash equilibrium point solution concept for the dynamic oligopoly advertising game with payoffs $E[U(\pi_j)]$. We say that $(d_1^*,\ldots,d_J^*)$ is a *discounted equilibrium point* (**ep**) with respect to H if

$$V_j(d_1^*,\ldots,d_j^*,\ldots,d_J^*\,|\,h) \geq V_j(d_1^*,\ldots,d_{j-1}^*,d_j d_{j+1}^*,\ldots,d_J^*\,|\,h) \qquad (6)$$

for all d_j, $j = 1,\ldots,J$ and $h \in H$.

3 The Risk-Sensitive Market Share Attraction Game

Let s_t, the *state of the sequential game at time* t, be the vector of goodwill levels *before* advertising decisions are made in period t. Also, define g_{jt} as firm j's action in period t. Thus the vector $g_t = (g_{jt})$ is the vector of goodwills *after* advertising decisions are made in period t but before profits are earned. From (1),

$$g_t = s_t + La_t \quad \text{and} \quad s_{t+1} = K_t g_t \qquad (7)$$

specify actions and states, respectively.

As in M-S, let m_j be the j-th row of $M = L^{-1}$. From (7), $a_t = M(g_t - s_t)$ and

$$a_{jt} = m_j(g_t - s_t). \qquad (8)$$

It is apparent from (8) that the constraint on nonnegativity of advertising expenditures is a *joint* constraint on the firms' selections of their goodwill levels! Although static games with joint constraints have been studied, in Section 5 we concentrate on the case where L is a diagonal matrix. In this case, if the diagonal elements of L are nonnegative, nonnegativity of a_{jt} corresponds to $g_{jt} \geq s_{jt}$. We use the expression for the advertising level for firm j in period t given in (8) to express net profit solely in terms of period

t's state and action vectors:

$$X_{jt} = \mu_j(\mathbf{g}_t) - a_{jt} = \mu_j(\mathbf{g}_t) - \mathbf{m}_j(\mathbf{g}_t - \mathbf{s}_t). \tag{9}$$

The equilibrium point problem is to find strategies $d_1^*, \ldots, d_J^*$ for the J firms that satisfy (6), where firm j's payoff is

$$\mathrm{E}\left\{ \exp\left(-\lambda \sum_{t=1}^{\infty} (\delta_j)^{t-1} X_{jt} \right) \right\}, \tag{10}$$

where X_{jt} is given in (9), subject to the dynamics in (7).

4 An Equivalent Static Game

In this section, we relate the dynamic stochastic game developed in the previous section to a family of *static* (i.e., single-period) deterministic games. This relationship leads to sufficient conditions for the dynamic game to have a *myopic* solution. That is, the equilibrium points of the one-period games comprise an equilibrium point of the dynamic game. To simplify the notation, we now place some restrictions on some of the parameters of the problem. Much of the analysis applies to a more general structure, however. Suppose that $\mathbf{K}_t$ is diagonal with θ_j in the j-th row and column, and $P\{0 \le \theta_j \le 1\} = 1$ for each j. Suppose also that $\mathbf{L}$ is diagonal with ℓ_j in the j-th row and column. Then $\mathbf{m}_j$, the j-th row of $\mathbf{L}^{-1}$, is a row vector with $1/\ell_j$ in the j-th position and zeros elsewhere. Under these assumptions, consider a static game wherein the payoff function for firm j has the following form:

$$v_j(\mathbf{g}, \alpha) = \mathrm{E}\left\{ \exp\left(-\alpha \left[\mu_j(\mathbf{g}) - \frac{1 - \delta\theta_j}{\ell_j} g_j \right] \right) \right\}, \tag{11}$$

where $\mathbf{g}$ is a J-vector of goodwill levels with the j-th component g_j.

Let $\theta_{j1}, \theta_{j2}, \ldots$ be independent and identically distributed random variables for each j with the same distribution as θ_j and independent of θ_j. Drawing on results in Shinde and Sobel (1991), the substitution of (7)–(9) in (10) yields

$$\mathrm{E}\left\{ \exp\left(-\lambda \sum_{t=1}^{\infty} (\delta_j)^{t-1} X_{jt} \right) \right\} = e^{\lambda M_j s_{j1}} \mathrm{E}\left[\prod_{t=1}^{\infty} v_j\left(\mathbf{g}_t, \lambda (\delta_j)^{t-1} \right) \right]. \tag{12}$$

This representation of each firm's criterion leads to the analysis of a single-period game with payoffs $v_j\left(\mathbf{g}, \lambda (\delta_j)^{t-1} \right)$.

Let $\Gamma(\alpha)$ be the noncooperative game among the J competitors in which player j chooses $g_j \ge 0$, $\mathbf{g} = (g_1, \ldots, g_J)$, and player j's payoff is $v_j(\mathbf{g}, \alpha)$ defined in (11). Suppose $\mathbf{g}(\alpha) = [g_1(\alpha), \ldots, g_J(\alpha)]$ is an unrandomized equilibrium point of $\Gamma(\alpha)$ and the *repeatability* condition

$$g_j\left(\lambda (\delta_j)^{t-1} \right) \le g_j\left(\lambda (\delta_j)^{t} \right) \quad \text{for each } j \text{ and } t \tag{13}$$

is satisfied. This condition guarantees that if firm j employs an equilibrium strategy $g_j\left(\lambda(\delta_j)^{t-1}\right)$ in period $t-1$, then it is *feasible* for firm j to implement the equilibrium strategy $g_j\left(\lambda(\delta_j)^t\right)$ in period t.

Let $\mathcal{S} = [0, g_1(\lambda)] \times \cdots \times [0, g_J(\lambda)]$ denote the J-dimensional rectangle of points whose components do not exceed $\mathbf{g}(\lambda)$. The next result follows from Shinde and Sobel (1991).

Lemma 1. *If (13) holds, then an equilibrium point of the dynamic risk-sensitive game, with respect to $\mathcal{S}$, is given by $g_{jt} = g_j\left(\lambda(\delta_j)^{t-1}\right)$ for all j and t. At this equilibrium point, player j's criterion (10) and (12) achieves the value*

$$\exp\left(\lambda M_j s_{j1}\right) \prod_{t=1}^{\infty} v_j\left(\mathbf{g}\left[\lambda\left(\delta_j\right)^{t-1}\right], \lambda\left(\delta_j\right)^{t-1}\right).$$

The challenge to applying Lemma 1 is confirming that (13) is satisfied. Since $\lambda\left(\delta_j\right)^{t-1} \to 0$ as $t \to \infty$, it would follow from (13) that risk-sensitive goodwill levels are lower than risk-neutral goodwill levels.

In the next section, we analyze $\Gamma(\alpha)$, focussing on the effects of α, hence risk-sensitivity, on its properties.

5 Risk-Sensitive Equilibrium Points

It is convenient to rewrite (11), firm j's payoff in the static game, as

$$v_j(\mathbf{g}, \alpha) = \exp[-\rho_j(\mathbf{g})\alpha]\phi_j\left(\frac{\alpha\delta g_j}{\ell_j}\right), \tag{14}$$

where

$$\phi_j(s) \equiv \mathrm{E}\left(e^{s\theta_j}\right), \tag{15}$$

$$\rho_j(\mathbf{g}) \equiv \mu_j(\mathbf{g}) - g_j/\ell_j, \tag{16}$$

and where $\mu_j(\mathbf{g})$ has the market share attraction form given in (3).

5.1 Symmetry

Let F_j be the distribution of θ_j and let

$$v_j^{(j)}(\mathbf{g}, \alpha) \equiv \frac{\partial v_j(\mathbf{g}, \alpha)}{\partial g_j}.$$

We examine the special case where all firms are symmetric; i.e., $b_j = b$, $M_j = M$, $\ell_j = \ell$, $F_j = F$, and $\delta_j = \delta$ for all j. Let $\phi(s) = \phi_j(s)$ in this symmetric case. Suppose we have a symmetric solution to the static game: $g_j^* = y$ for all j. Let $y = y^\alpha$ satisfy the first-order condition for each player's optimization problem. We now establish the fact that y^α is a decreasing function of α.

43

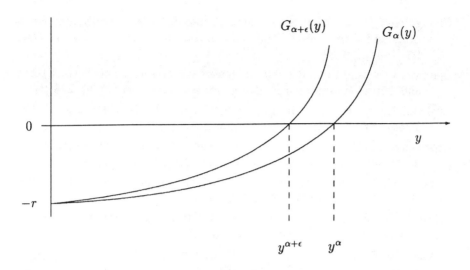

Figure 1: First-order conditions for α and $\alpha + \epsilon$.

Theorem 1. *In the symmetric game, the equilibrium goodwill level is a decreasing function of α.*

Sketch. Write the generic first-order condition as $G_\alpha(y) = 0$. The result follows from this observation: if $\partial G_\alpha(y^\alpha)/\partial y > 0$ and $\partial G_\alpha(y)/\partial \alpha|_{y=y^\alpha} \geq 0$, then $y^\alpha \geq y^{\alpha+\epsilon}$ for $\epsilon > 0$ sufficiently small. See Figure 1. □

Since $\alpha = \lambda(\delta)^{t-1}$, the form of the dependence of the equilibrium goodwill levels on λ, δ, and t is immediate.

Corollary 1. *In the symmetric game, equilibrium goodwill levels are decreasing in each of the following parameters: the risk parameter λ, the single-period discount factor δ, and the time period t.*

5.1.1 Interpretation

Let y_t be a solution to $G_\alpha(y) = 0$, when $\alpha = \lambda(\delta)^{t-1}$. In the symmetric game, a symmetric equilibrium point consists of $g_{jt} = y_t$ for each j and t. It follows from Theorem 1 that $y_1 \leq y_2 \leq \cdots$.

References

Bell, D. E., "Risk, Return, and Utility," *Management Sci.*, **41** (1995), 23–30.

Bouakiz, M. and M. J. Sobel, "Inventory Control with an Exponential Utility Criterion," *Operations Research*, **40** (1992), 603–608.

44

Chung, K.-J., and M. J. Sobel, "Discounted MDP's: Distribution Functions and Exponential Utility Maximization," *SIAM J. Control Opt.*, **25** (1987), 49–62.

Denardo, E. V. and U. G. Rothblum, "Optimal Stopping, Exponential Utility, and Linear Programming," *Math. Programming*, **16** (1979), 228–244.

Eagle, J. N., "A Utility Criterion for the Markov Decision Process," Ph.D. Thesis, Stanford University, Stanford, CA (1975).

Eeckhoudt, L., C. Gollier, and H. Schlesinger, "The Risk-Averse (and Prudent) Newsboy," *Management Sci.*, **41** (1995), 786–794.

Erickson, G. "Empirical Analysis of Closed-Loop Duopoly Advertising Strategies," *Management Sci.*, **38** (1992), 1732–1749.

Erickson, G. "Differential Game Models of Advertising Competition," *European J. of Oper. Res.*, **83** (1995), 431–438.

Howard, R. S. and J. E. Matheson, "Risk-Sensitive Markov Decision Processes," *Management Sci.*, **18** (1972), 356–369.

Jaquette, S. C. "A Utility Criterion for the Markov Decision Process," *Management Sci.*, **23** (1975), 43–49.

Jørgensen, S. "A Survey of Some Differential Games in Advertising," *J. of Econ. Dynamics and Control*, 4 (1982), 341–369.

Monahan, G. E. and M. J. Sobel, "Stochastic Dynamic Market Share Attraction Games," *Games and Econ. Behavior*, **6** (1994), 130–149.

Nerlove, M. and K. Arrow, "Optimal Advertising Policy Under Dynamic Conditions," *Economica*, **22** (1962), 129–142.

Porteus, E. L. "On the Optimality of Structure Policies in Countable Stage Decision Processes," *Management Sci.*, **17** (1975), 411–426.

Shinde, M. and M. J. Sobel, "Myopic Equilibrium Points in Risk-Sensitive Dynamic Oligopoly," unpublished manuscript, SUNY Stony Brook, Stony Brook, NY (1991).

Sorger, G. "Competitive Dynamic Advertising," *J. of Econ. Dynamics and Control*, **13** (1989), 55–80.

Whittle, P. "Risk-Sensitive Linear/Quadratic Gaussian Control," *Adv. Applied Probability*, **13** (1981), 764–777.

Specification and Estimation of Nonlinear Models with Dynamic Reference Prices

Harald Hruschka[1] and Martin Natter[2]

[1] University of Regensburg, Germany
[2] Vienna University of Economics, Austria

Abstract: We specify reference prices as nonlinear dynamic latent variables represented as hidden units of an artificial neural net. Model parameters are estimated by an extended version of backpropagation. For a scanner data base containing prices and sales of seven brands of a consumer non-durable these models lead to better fits as compared to models that conceive reference prices as expectations. All reference price models are tested against a basic model with various functional forms. Results emphasize the importance of reference prices for the explanation of market share response.

1 Introduction

Reference price theory provides a behavioral explanation of dynamic effects on brand choice and derived aggregated market response measures caused by prices of the past. It is assumed that consumers store price information gathered on previous purchase occasions. Consumers use this internal price (Tull et al. 1964) as standard of comparison or reference to evaluate observed prices (Winer 1988). If the reference price is higher than the current price, consumers judge the later favorably. If, on the other hand, the reference price is lower, the current price is judged unfavorably. Favorable (unfavorable) judgment of a current price increases (decreases) market share. From this it follows that reference prices play a role in shaping optimal dynamic pricing strategy which is, after all, the reason for the practical relevance of the concept under discussion (Natter 1994,Greenleaf 1995).

This work addresses the question of how to specify and measure reference prices. Predictors have hypothetical effects on reference prices, these on their turn influence market shares. Three types of reference price mechanisms are investigated: extrapolative expectations, rational expectations and dynamic latent variables.

2 Basic Model Without Reference Prices

A market share model using relative price of the considered brand, relative prices of competing brands and lagged market share as predictors serves as benchmark for the reference price models described later:

$$S_{i,t} = f(z_{i,t}) = f(a_{i,0} + \sum_j a_{i,j} p_{j,t} + a_{i,J+1} S_{i,t-1}) \qquad (1)$$

$S_{i,t}$ denotes market share of brand i in period t, $p_{i,t}$ its relative price defined as ratio of actual consumer price and average price over all brands considered.

Market share of brand i decreases with higher relative prices, because in this case many consumers switch to other brands. Therefore one expects negative values of a_{ii}.

Since the brands concerned belong to one product category, we expect substitutional effects. The signs of the parameters measuring effects of the prices of competing brands should therefore be positive, because as a rule demand of the brand under study rises if the price of another brand increases and other conditions stay the same. As relative prices of all brands always sum up to J, each relative price can be represented as a linear combination of the other relative prices. We therefore include only those relative prices in each market share equation that possess coefficients significantly different from zero.

Customer holdover effects $a_{i,J+1} S_{i,t-1}$ represent the portion of the demand caused indirectly by past marketing decisions that lead to consumer satisfaction or positive word-of-mouth.

$f(.)$ represents the functional connection between predictors and market shares. Out of the various frequently used functional forms of market response models (Hanssens and Parsons 1993, Lilien et al. 1992) besides the linear model we confine ourselves to the exponential and the logistic.

3 Reference Prices as Extrapolative or Rational Expectations

According to the concept of extrapolative expectations consumers assume that price development follows an univariate time series model (e.g. an autoregressive model). On the other hand, proponents of the concept of rational expectations (Muth 1961) suggest that all information available should be used to explain formation of expected prices.

Consumers' price expectation are not directly measurable. Consequently expectations are simply equated with prices estimated on the basis of a model with price as dependent variable. Therefore models with extrapolative or rational expectations consist typically of two equations. The first equation describes the price expected by consumers. The second equation explains market share among other things by surprise effects caused by a difference between current price and expected price generated by the first equation.

Equations EE1 up to EE4 all follow the concept of extrapolative expectations. In EE1 we examine the time horizon of past price influences on price expectations with a maximum lag of five periods. EE2 and EE3 assume a first and second order autoregressive process of price formation, respectively. EE4 adds a time trend to a first order autoregressive process.

The rational expectations equation RE serves to test a number of hypothetical effects on expected price stemming from variables that are lagged one period:

- price;
- prices of competing brands;
- market share (positive):
- sales volume of the respective outlet (positive);

as well as a linear time trend.

We consider three different types of market share equations (see also table 1):

- **MS1:** Market share increases with the difference between reference price and observed price ($a_{i,3}$ positive).
- **MS2:** Extends MS1 by adding reference prices of competing brands. Higher differences between reference price and observed price for a competing brand j motivates more consumers to buy this brand, which leads to a lower market share of brand i ($a_{i,J+1+j}$ negative).
- **MS3:** Considers asymmetric effects of positive and negative differences between reference prices and observed prices.

A theoretical basis is provided by Prospect Theory (Kahneman and Tversky 1979), according to which consumers respond more strongly to losses as compared to gains relative to a specific frame of reference (here: a reference price). Therefore the absolute value of the coefficient for negative differences between a brand's own reference prices and observed prices (higher observed prices), which are losses to the consumer, should be higher than the absolute value of the coefficient for positive differences. Conditions are, of course, the other way round for prices of competing brands.

Asymmetric effects contradicting Prospect Theory (i.e. prices below the reference level are more important) could be caused by the fact that price cut effects are frequently intensified by feature advertising (Simon 1989). Other possible explanations are brand switching costs which motivate consumers to benefit from lower prices of prefered brands (Bultez 1975) or a high proportion of brand switchers attracted by lower prices (Krishnamurthi et al. 1992).

Table 1: Extrapolative or Rational Expectations: Market Share Equations

	Equations
MS1	$S_{i,t} = f(a_{i,0} + a_{i,1}p_{i,t} + a_{i,2}S_{i,t-1} + a_{i,3}(p_{i,t}^r - p_{i,t}) + \epsilon_{i,t})$
MS2	$S_{i,t} = f(a_{i,0} + \sum_j a_{i,j}p_{j,t} + a_{i,J+1}S_{i,t-1} + \sum_j a_{i,J+1+j}(p_{j,t}^r - p_{j,t}) + \epsilon_{i,t})$
MS3	$S_{i,t} = f(a_{i,0} + \sum_j a_{i,j}p_{j,t} + a_{i,J+1}S_{i,t-1} +$
	$\quad \sum_j a_{i,J+1+j}(p_{j,t}^r - p_{j,t})^- + \sum_j a_{i,2J+1+j}(p_{j,t}^r - p_{j,t})^+ + \epsilon_{i,t})$

4 Reference Prices as Dynamic Latent Variables

In the sequel, we examine models with dynamic reference prices. The model class addressed here contains three types of variables: predictors, reference prices and market shares. Predictors are price, prices of competing brands, lagged market share, and trend. The reference price equations show the effects of predictors on reference prices, the market share equations the effects of reference prices and predictors on market share. We replace relative prices by absolute prices here, because they lead to better fits for the models with dynamic latent variables.

The models of this section form reference prices as latent variables in such a way that variance explanation of market share is maximized. This differs from the expectation models discussed above, where price equations maximize variance explanation of price. This approach could result in models with relativley low variance explanation of market shares, especially if consumers use only part of the available information to form their internal prices.

We represent long term influences of the consumers' internal price formation by introducing autoregressive effects of these latent variables, i.e. the reference prices (Elman 1991).

The first model variant, DM1, considers two types of reference prices. Every consumer has a reference price for every brand but, in addition, he considers an average reference price that is equal for all brands, but whose influence on other brands can vary. We assume that this common average reference price is influenced by the average price of the preceding period, the sales volume of the preceding period, a linear trend and autoregressive effects of the latent variable up to the second order.

The entire model can be represented in the form of an equation system (see Table 2).

Table 2: Equations of the Models with Latent Variables

Model	Equation
DM1	$S_{i,t} = f(a_{i,0} + a_{i,1}p_{i,t}^r + a_{i,2}\bar{p}_t^r + \sum_j a_{i,j+2}p_{j,t} + a_{i,J+3}S_{i,t-1} + a_{i,J+4}t + \epsilon_{i,t})$
	where:
	$p_{i,t}^r = f(b_{i,0} + b_{i,1}p_{i,t-1} + b_{i,2}Q_{i,t-1} + b_{i,3}t + b_{i,4}p_{i,t-1}^r + b_{i,5}p_{i,t-2}^r)$
	$\bar{p}_t^r = f(\beta_{i,0} + \beta_{i,1} \sum_j p_{j,t-1} + \beta_{i,2}Q_{t-1} + \beta_{i,3}t + \beta_{i,4}\bar{p}_{t-1}^r + \beta_{i,5}\bar{p}_{t-2}^r)$
DM2	$S_{i,t} = f(a_{i,0} + \sum_j a_{i,j}p_{j,t}^r + a_{i,J+1}\bar{p}_t^r + \sum_j a_{i,J+1+j}p_{j,t} + a_{i,2J+2}S_{i,t-1}$
	$+a_{i,2J+3}t + \epsilon_{i,t})$
	where:
	$p_{i,t}^r = f(b_{i,0} + b_{i,1}p_{i,t-1} + b_{i,2}Q_{i,t-1} + b_{i,3}t + b_{i,4}p_{i,t-1}^r + b_{i,5}p_{i,t-2}^r)$
	$\bar{p}_t^r = f(\beta_{i,0} + \beta_{i,1} \sum_j p_{j,t-1} + \beta_{i,2}Q_{t-1} + \beta_{i,3}t + \beta_{i,4}\bar{p}_{t-1}^r + \beta_{i,5}\bar{p}_{t-2}^r)$

In table 2 Q_t symbolizes the total sales volume of an outlet in period t, $\bar{p}_t^r$ the average reference price of an outlet in Period t, $Q_{i,t}$ the sales volume of brand i in period t, $p_{i,t}$ its price and $p_{i,t}^r$ its reference price.

Since the comparative equations MS2 and MS3 include also reference prices of competing brands, we specify model DM2 as an extension of model DM1 which allows effects of reference prices of other brands.

We postulate that higher observed prices in previous periods increase reference prices. The effect of a brand's own reference price on market share should be positive. Market share increases for a given actual price, if the own reference price becomes higher, as this renders this actual price less harmful in the eyes of the consumers. On the other hand, the effect of reference prices of competing brands on market share should obviously be negative.

5 Estimation and Model Selection

In this work artificial multilayer neural networks are used to specify and measure reference prices. Applications of artificial neural networks to the determination of market response models may be found in Hruschka (1993) or Hill et al. (1994). Networks for models with extrapolative or rational price expectations consist of two layers, one for the predictors, the other for market shares and prices of the brands as output variables. Networks for models with dynamic latent variables possess three layers, for predictors, latent variables (hidden units) and market shares as output variables, respectively. The layer with latent variables contains the reference prices. Feedbacks of reference prices serve to reproduce effects of past reference prices on current reference prices. Multilayer networks are capable to approximate any continuous function up to any small error (Cybenko 1989,Hornik et al. 1989,Ripley 1993).

We determine model parameters by a variant of backpropagation, which is the most widespread estimation method for multilayer neural networks. For the models with extrapolative or rational expectations backpropagation leads to results equal to OLS-estimation. Model selection is made by a backward selection procedure where estimation, diagnostic and model simplification stages alternate. The null hypothesis that a model parameter β equals zero is examined by an appropriate F-test (Judge et al. 1980) assuming that residuals are asymptotically normally distributed and using a procedure of Bishop (1992) to compute the Hessian for multilayer networks.

As badness-of-fit measure we use the prediction criterion (PC) developed by Amemiya (1983) on the basis of the sum of squared residuals (SSE). It punishes a higher number of parameters (P) more severely than the corrected coefficient of determination known from multiple linear regression (N denotes the number of observations):

$$PC = \frac{SSE}{N - P}(1 + \frac{P}{N}) \tag{2}$$

6 Empirical Study

6.1 Database and Pooling

The empirical study is based on scanner data consisting of time series of sales and prices for 7 brands of a consumer nondurable acquired in 21 retail outlets over a time span of 73 weeks. We use Ward's hierarchical clustering algorithm to group outlets on the basis of the parameters of a simple linear model. Likelihood Ratio Tests allow us to proceed with pooled models.

6.2 Results of the Basic Model

The best fit is obtained by using the logistic functional form. The exponential form usually shows only slight disadvantages. Both nonlinear forms clearly outperform the linear form with regard to PC. The logistic model appears best suited as standard for the reference price models. With one exception all signs of the coefficients measuring the effects of prices of competing brands are positive. With a mean variance explanation of $R^2 = 0.447$ and $PC = 496$ (PC-values are multiplied by 10^5) this basic model prevails over its related rivals.

6.3 Results of Models with Extrapolative or Rational Expectations

Estimation of price expectation equations with autoregressive terms up to a maximum lag of 5 weeks rarely produces more than 2 significant lags. Restricting the equations this way leads to an average variance explanation of 26.8%. The coefficient of the price lagged by one period is always positive and significant, for the prices lagged by 2 periods 19 out of 28 coefficients are significant (2 thereof are negative). If prices lagged by two periods are omitted and price is predicted only on the basis of the price of the previous period (equation EE2), then the explanation changes slightly for the worse ($R^2 = 24.6\%$) with all parameters being significant. The incorporation of prices lagged by two periods results in a modest improvement of the variance explanation of the expected price, but the change in market share explanation is practically equal to zero.

The two extrapolative price equations EE3 and EE4 have the same number of parameters and the same variance explanation ($R^2 = 0.268$). Equation EE4 has significant coefficients for lagged own price and time trend. If the expected prices of equation EE4 are inserted into the linear market response function MS1, then this function shows a slight increase to $R^2 = 0.431$. Despite the incorporation of additional predictors, the rational expectation equation RE shows only slight improvements. We therefore use EE4 as price equation to compute expected prices. Since among the basic models the logistic model outperformed the exponential, in the sequel we refrain from estimating the exponential version.

Generally, one can say that models with asymmetric reference prices ($R^2 = 0.525$) achieve a better fit than the symmetric ($R^2 = 0.466$) models. The same goes for the logistic models vis-a-vis their linear counterparts. The asymmetric logistic model (MS3 - logistic) attains, on average, a variance explanation as high as $R^2 = 0.557$ and a PC of 407, the best one so far obtained. Contradicting Prospect Theory, absolute coefficients for gains (observed prices smaller than reference prices) are higher.

6.4 Results of Models with Dynamic Latent Variables

Model DM1 explains 55.7% of the variance on average, its PC-value is 409. Model DM2 that adds reference prices of competing brands achieves a higher mean variance explanation of 58.3% and a prediction criterion of 393. Therefore this model is superior to models without latent variables. The comparatively poor result when using an average reference price and a brand's own reference price suggests that a brand competes usually with one or two other brands of the product category concerned.

Before applying the automated model selection procedure each market share model has 17 parameters. This number is reduced to an average of 9.4 significant parameters. A brands' own reference price is significant in 89.2%, the average reference price in 42.8% of the equations.

The signs of the effects of previous period's price on reference prices and of own reference prices on market share are on the whole positive. This conforms to our hypotheses. The hypothesis that lagged sales volume of an outlet affects the reference price is to be rejected. Trend effects, on the other hand, occur frequently, but their impact is not uniform. One-period lags of reference prices are significant in 50% , two-period lags in 32% of the estimated equations. In most equations there is a positive connection between reference prices and their autoregressive predecessors, demonstrating some inertia in reference price formation.

7 Conclusion

This paper presents a new approach to reference price modeling. Using multilayer neural network models, reference prices are specified and estimated by backpropagation. The results achieved show an improvement over similar models without reference prices.

The research described yielded the following results:

- Logistic functions give better model fits compared to exponential and linear forms.

- Contrary to Prospect Theory reference prices above observed prices (Gains) have stronger effects than lower reference prices (Losses).

- Models with reference prices are superior to models without reference prices. On the average a model, which allows asymmetric effects of prices of competing brands, increases the variance explained of the basic model by 11 percentage points. A latent variable model increases the variance explained by 13.6 percentage points.

- Reference prices of competing brands are important. Customers obviously take prices of two or three competing brands into consideration.

References

Amemiya, T. (1980), "Selection of Regressors," *International Economic Review*, 331–354.

Bishop, C. (1992), "Exact Calculation of the Hessian Matrix for the Multilayer Perceptron," *Neural Computation*, 494–501.

Bultez, A. (1975), "Price Cut versus Coupon Promotion", Working Paper, European Institute for Advanced Studies in Management, Brussels.

Cybenko, G. (1989), "Continuous Value Neural Networks with Two Hidden Layers are Sufficient," *Mathematics of Control, Signal and Systems*, 303–314.

Elman, J.I. (1991), "Distributed Representations, Simple Recurrent Networks, and Grammatical Structure," *Machine Learning*, 7, 195–225.

Greenleaf, E. (1995), "The Impact of Reference Price Effects on the Profitability of Price Promotions, *Marketing Science*, 82–104.

Hanssens, D.M. and L.M. Parsons, (1993), "Econometric and Time-Series Market Response Models," in *Marketing*, Eliashberg, J. and G.L. Lilien, Eds., Amsterdam: North-Holland.

Hill, T., Marquez, L., O'Connor, M. and W. Remus (1994), "Artificial Neural Network Models for Forecasting and Decision Making," *International Journal of Forecasting*, 5–15.

Hornik, K., Stinchcombe, M. and H. White (1989), "Multilayer Feedforward Networks are Universal Approximators," *Neural Networks*, 2, 359–366.

Hruschka, H. (1993), "Determining Market Response Functions by Neural Network Modeling. A Comparison to Econometric Techniques," *European Journal of Operational Research*, 66, 27–35.

Judge, G.E., Griffiths, W.E, Hill, R.C and T.C. Lee (1980), *The Theory and Practice of Econometrics*, New York: Wiley.

Kahneman, D. and A. Tversky, "An Analysis of Decision Under Risk," *Econometrica*, S. 363–391.

Krishnamurthi, K., Mazumdar, T. and S.P. Raj (1992), "Asymmetric Response to Price in Consumer Brand Choice and Purchase Quantity Decisions," *Journal of Consumer Research*, 387–400.

Lilien, G.L., Kotler, P. and K.S. Moorthy (1992), *Marketing Models*, Englewood Cliffs, NJ.: Prentice-Hall.

Muth, J.F. (1961), "Rational Expectations and the Theory of Price Movement," *Econometrica*, 315–335.

Natter, M. (1994), "Nichtlineare Marktreaktion und Stochastisches Preismanagement im Mehrproduktunternehmen: Ein Konnexionistischer Ansatz," Ph.D. Thesis, Vienna, Austria: University of Economics.

Ripley, B.D. (1993), "Statistical Aspects of Neural Networks, in *Networks and Chaos - Statistical and Probabilistic Aspects*, Barndorff-Nielsen, O.E., Jensen, J.L. and W.S. Kendall, Eds., London: Chapman & Hall.

Simon, H. (1989), *Price Management*, North Holland: New York.

Tull, D.S., Boring, R.A. and M.H. Gonsior (1964), "A Note on the Relationship of Price and Imputed Quality," *Journal of Business*, 180–191.

Winer, R. (1988), "Behavioral Perspective on Pricing: Buyers' Subjective Perceptions of Price Revisited," in *Issues in Pricing*, Devinney, T., Ed., Lexington, MA: Lexington Books.

Asymmetric Dynamic Switching between High and Low Quality Brands: Empirical Evidence from the US Car Market[1]

Chung Koo Kim[1]

[1] Department of Marketing, Concordia University, Montreal, Quebec H3G 1M8, Canada

Abstract. Although *asymmetric switching* due to price deals between high quality and low quality brands is well documented, the *dynamic* aspect of asymmetry has not been explored. This paper empirically examines the asymmetric lead and lag effects of short-run price decisions on the market shares across different quality brands in the US small car market.

Key words. Competition, quality, price effect, asymmetric switching, automobiles

1. Introduction

How do people respond to price cuts or price promotions for high-quality image brands? Do they respond in the same way to price cuts for low quality brands? People seem to show different response behaviours for the two different cases; Generally they tend to be more responsive to a price cut for high-quality brands than for low-quality brands. As a result, sales tend to increase significantly more for high-quality brands than for low-quality brands when prices are reduced in the same proportion (e.g., 20%). The *asymmetric sales effect* of price promotions favouring high-quality brands has been an important topic for both marketing practitioners and academics.

Another important issue relating to price promotions has been its dynamic effect on sales or shares. A brand's price reduction in a certain period would have an effect on the future sales of both its own brand and the other brands. The phenomenon is called *lag effect.* Several time-series studies also showed that a brand's price decrease in a future time period significantly influences consumer demand in the current time period, when buyers could anticipate in advance the future price deals. For example, if consumers anticipate that *Ford Taurus* will provide buyers with a $2,000 cash-back rebate in the near future, some people will not buy in the current period and will wait

1. The author would like to thank Brian Ratchford at SUNY Buffalo, Subrata Sen at Yale, and Eunsang Yoon at University of Massachusetts Lowell for their helpful comments on the previous draft, which was presented at the 1993 Marketing Science Conference at St. Louis. He also acknowledges that this research paper is supported in part by a research grant from Social Sciences and Humanities Research Council of Canada (SSHRC).

until later. Therefore, the sales of *Ford Taurus* and the other brands in the current period can be influenced by consumer anticipation. This phenomenon is called *lead or anticipation effect*.

Will a price promotion for high-quality brands have stronger lead and lag effects than for low-quality brands? Car market data shows that if people anticipate that a brand (especially a high-quality brand) will provide a price deal through a cash rebate, then the sales of other brands will drop in the current period because of the future deal expectation. Low-quality brands are especially vulnerable to the other competitors' future rebate programs (*Automotive News 1982-87*).

This paper examines the *asymmetric* effects between high and low quality brands in the *dynamic* context. The primary focus is on testing the *asymmetric lead and lag* effects of short-run price decisions (price cut or promotion) on the shares between brands. This paper is different from the existing studies of *asymmetry*, because it addresses the issues in the *dynamic* context and for the case of a durable (the subcompact car market). A multiple time-series method is employed to examine the dynamic nature of asymmetric effects.

2 Literature and Hypotheses

2.1 The Literature on Asymmetric Effect

Asymmetric switching or share changes between high quality and low quality brands, has been well documented. Empirical studies have shown that price reductions for higher quality brands attract more buyers than do price reductions for lower quality brands (Allenby and Rossi 1991; Blattberg and Wisniewski 1989; Carpenter et al. 1988; Kamakura and Russell 1989). When a high quality brand price deals, it steals sales away from both other high quality brands and low quality brands. However, a low quality brand takes unit sales from other low quality brands, but rarely takes sales away from high quality brands.

Carpenter et al. (1988) found that economy brands are vulnerable to each other's price cuts, but have little impact on other brands. Premium brands are affected the least by price competition from economy brands. Asymmetries in competition were attributed to two main sources; (1) unique feature of brand strategy, e.g., unique distribution, reputation, etc., and (2) period-to-period variation in marketing-mix elements.

Blattberg and Wisniewski (1989) also documented that when higher-price brands price deal, they steal sales away from their own tier (the same-quality group) and the tier below (the lower quality group). The study provides a statistical explanation with preference distribution.

Kamakura and Russell (1989) empirically showed that a price cut by private labels has little impact on the sales of national brands, but the sales of private labels are strongly affected by a price change by national brands. No specific explanation was provided but they attributed the asymmetry to the price tier theory of Blattberg and Wisniewski (1989).

Allenby and Rossi (1991) also documented that a higher quality brand has a higher price elasticity than does a lower quality brand with the same market share. The asymmetry between high and low quality brands was attributed to the interaction between income and substitution effects. As a superior brand lowers its price, it induces a substitution away from lower quality brands and an income effect in the same direction. However, when an inferior brand lowers its price, substitution from the higher quality brands will be induced, but the income effect will work in the opposite direction.

Some noticeable common features of those studies are:

1. They have focused on *static* estimates of price strategies, although in many cases *dynamics* are important, as mentioned in Carpenter et al. (1988).

2. They have addressed the asymmetric switching mainly for the cases of frequently purchased grocery products.

Therefore, it has not yet been documented whether or not the theory of asymmetry could extend to the dynamic case. It would also be quite interesting to examine the asymmetric switching for the case of non-frequently purchased durables such as cars.

2.2 The Literature on Lead and Lag Effect

There has been another stream of marketing studies on *dynamic* marketing effects on market sales or shares in the short run. Current expenditures on marketing instruments usually do not have their full impact on sales in the same period. Moreover, their impact on sales may extend well into the future. This is called the *lag* (or *carryover*) effect. It is also possible that consumers anticipate a marketing action and adjust their responses even before the action takes place. Therefore, sales may react to marketing efforts in the future. This is called the *lead* (or *anticipation*) effect (Doyle and Saunders 1985; Hanssens, Parsons, and Schultz 1990).

The lead and lag effects of marketing variables has been examined by several time-series studies (Aaker, Carman, Jacobson 1982; Bass and Pilon 1980; Doyle and Saunders 1985; Hanssens 1980; Jacobson and Nicosia 1981; Leone 1983). The lead or lag effects may exist not only within a brand but also between brands or even between different sub-markets (e.g., different quality groups). In other words, when people anticipate that a brand will provide an incentive program in the near future, the sales of other brands may decline in the current period.

The existing studies have examined the lead and lag effects mainly within a product/brand but not between competing products/brands. One exception is Hanssens' (1980) in which *cross*-lag effects of a brand's marketing variables on others' sales were examined, although the cross-lead effect was not addressed.

2.3 Objectives and Major Hypotheses

The major objective of the current study is to provide an empirical evidence of asymmetric switching between brands in the dynamic context. This study examine how a brand's price cut or price promotion has the asymmetric lead and lag effects on the market shares of other brands. More specifically, two major hypotheses are developed and empirically tested using the car market data.

First, as an extension of the existing time-series studies, the current study shows that *cross*-lag and *cross*-lead effects exist among the competing brands in a market. Existing time-series study showed that a brand's price cut in the current period would have a negative effect on the shares of other brands. It was also shown that rebates and low financing offers showed some significant lead or anticipation effects on other competing brands. In other words, when people anticipated that a brand (especially of good quality) would provide a rebate offer in the near future, then the sales of other brands significantly dropped in the current periods. Therefore we have the following hypothesis:

H_1: Short-run price decisions (price cuts or promotions) of some brands have significant cross-lead as well as cross-lag effects on the market shares of the other brands.

This first hypothesis is consistent with existing time-series studies and the institutional fact of the automobile industry and the hypothesis will be used as a basis to further address asymmetric dynamic switching (*Automotive News* 1982-1987; Doyle and Saunders 1985; Hanssens 1980).

Second, by relating the time-series issues of lead and lag effects to those of asymmetry, it could be argued that high quality brands can have significant lag or lead effects on the other high-quality and low-quality brands while low quality brands will have less significant or insignificant lag or lead effects on high quality brands. This argument seems consistent with the existing studies of asymmetric switching, in which the asymmetric effect of marketing variables (especially price) has been well documented (Allenby and Rossi 1991; Blattberg and Wisniewski 1989; Carpenter et al. 1988; Kamakura and Russell 1989). In the real world, it has been shown that the shares of lower-priced or lower-quality brands tended to be more significantly affected by both previous and future rebate programs of higher-quality brands (*Automotive News* 1982-87). Therefore we have the following hypothesis:

H_2: A price cut or price promotion for higher quality brands has significant lead and lag effects on the shares of lower quality brands, but the reverse is not true.

Since the existing studies have not looked at the *asymmetric lead and lag effects*, empirical support from the present study will strengthen the theory of *asymmetric switching or share changes*. Also since no existing study provides a formal test for asymmetric share effects in durables such as cars, the results of this current project would contribute to generalizing the theory of asymmetric share effects of short-run price variables.

3 Method

3.1 A Time-series Methodology

To examine the *dynamic nature* of short-run marketing effects, the current study will employ a multiple time-series method. Although the time-series method is generally not new, it is selected because it provides a formal test procedure for the dynamic (i.e., lead and lag) effects of price cut or price promotion. More specifically, a *time series cross-correlation analysis* will be performed to examine the *asymmetric* competition effects among brands.

In order to formally analyze competitive interactions or reactions among brands, the following procedure will be employed, according to Hanssens (1980) and other time-series studies.

1. Develop univariate ARIMA models for major variables such as market shares and price variables, and save the white-noise ARIMA residuals.
2. Cross-correlate the ARIMA residuals with each other for the case of interest, i.e., short-run prices (cash rebate and low financing offer) and market shares.
3. Use Pierce-Haugh's chi-square test[2] for the significance of dynamic effects and inspect individual cross-correlations to examine the structure of cross-lead and lag effects.

Additional details about time series analysis can be obtained from most econometric and time-series books such as Judge et al. (1987), Maddala (1978), and Mills (1990). However, an excellent discussion in a marketing context is available in Hanssens, Parsons, and Schultz (1990).

3.2 Data Description

The data base used for this study is a time series and cross-section sample of demand sales, market shares, and marketing mix variables, for the period 1982-1987 in the sub-compact car market in the USA. The marketing mix variables included retail prices, two price promotion variables (i.e., rebates, low financing rates), and advertising expenditures for major brands in the subcompact car market.

Because this paper is interested in *asymmetry* between different quality brands, major brands[3] are categorized into three different quality groups (high, medium, and low quality groups) according to the measures of their *predicted reliability* evaluated by *Consumer Reports* (1 being much worse than average and 5 being much better than average). *Nova* and *Colt* were of high quality, *Escort* and *Lynx* medium quality, and the others lowest quality.

Ford Escort was the market leader although its predicted reliability was rated as

---

2. Pierce and Haugh's chi-square test was adopted for our analysis because of its ease of use and interpretation, as discussed in Hanssens (1980) and it was also used in several time-series studies including Aaker, Carmen, and Jacobson (1982).

3. Initially the data set included nine American brands and seven Non-American brands. On average, non-American brands were dominantly rated as high quality (except *Golf*) and tended not to give price promotion through rebates and lower financing rates. Because non-American brands rarely provided price cuts through consumer rebates and/or low financing offers during the study period, however, this paper cannot examine the effects of their price cuts on the shares of both American and non-American brands. Since an initial time-series analysis showed that price cuts for American brands did not have significant dynamic effects on non-American brands, it was decided that this paper focus on examining the dynamic effects among American brands.

medium. The *Ford Escort* was also active in marketing including advertising and price promotions through rebates and low financing offers. Among low-quality brands, *Horizon* and *Omni* were also active in marketing such variables as rebate programs, low financing offers, and advertising. A late entrant, *Nova*, could quickly increase its share and was active in terms of advertising, and price promotions.

Because of its importance in the asymmetric effect, the measure of quality is discussed in further detail, along with the measures of short-run prices and market share variables.

3.3 Measurement of Quality, Prices, and Market Share

In the existing studies of asymmetric competition, quality was indicated primarily by retail shelf price. The studies examined the asymmetric switching between "premium national brands," "economy brands," and "generics." It was shown that the different groups have significant price differentials. These separate distinct retail price groups, called *price tiers*, represented the quality differences (Blattberg and Wisniewski 1989; Carpenter et al. 1988; Kamakura and Russell 1989).

The quality variable is measured here as *predicted reliability* based on a model's Frequency-of-Repair history according to *Consumer Reports*. To derive these measurements, *Consumer Reports* compare each vehicle's reliability with the average reliability of all the models in their survey, and assign each brand to one of the five point scales; (1) much better than average, (2) better than average, (3) average, (4) worse than average, and (5) much worse than average. Major brands are accordingly categorized into high, medium, and low quality brand groups; *Nova* and *Colt* as the high-quality brands, *Escort, Lynx,* and *Alliance* as the medium-quality brands, and *Chevette, Horizon, Omni,* and *Pontiac 1000* as the low-quality brands.[4]

Because the cash rebate and the lower finance rate have been the two important methods of providing price deals for automobile buyers, this study focuses on examining the (dynamic) effects of the two variables on market share. The variable, *rebate*, is measured as the average amount of cash dollars (e.g., $500) given as a special price deal during a period (a month). The variable, *finance rate*, is measured as the average of the finance rate applied for each purchase during a month period. Sometimes, for example, a 2.9% finance rate was applied as a special price deal instead of a normal finance rate of 13.5% during the study period.

Market share is defined as the percentage share of a brand among all the brands in the car segment, measured as a brand's sales (the number of cars sold) divided by the total number of cars sold in the car market segment.

---

4. Several industry data sources indicate that reliability was regarded as more important than other measures such as performance, size, etc., when evaluating the quality of cars, and therefore frequently used by consumers as a proxy for product quality measure in the small car market (e.g., *Automotive News* 1982-87). Quality categorization based on *Consumer Reports'* measure of *predicted reliability* turned out to be equivalent to that based on *reliability factor* obtained from a factor analysis of major product attributes.

4 Empirical Application

A time-series cross-correlation analysis was performed according to the procedure discussed above. First, each time-series was prewhitened using the univariate ARIMA filters. The prewhitened residuals for each series were cross-correlated and chi-square statistics and two-tail test statistics were calculated to evaluate the significance of the lag and lead effects of the short-run price variables.

4.1 Univariate Analysis For Prewhitening

For the present study, in order to prewhiten each time series, first differencing to handle trends, and twelve differencing to deal with seasonality were tried. It was found that the above two methods produced no better results than no differencing. Therefore, it was decided to use natural logarithms of the data to handle nonstationarity in variance, i.e., heteroscedasticity.

Analysis of the patterns of autocorrelation functions (ACF) and partial autocorrelation functions (PACF) indicated that most series had followed ARIMA (1,0). The series which had not followed ARIMA (1,0) were prewhitened by applying other appropriate ARIMA filters.

4.2 Cross-correlation Analysis and Chi-square Test

In order to examine how the short-run pricing decisions (rebate and low financing offers) affect the brand shares of the other brands, the prewhitened series of brand shares were cross-correlated with those of the two short-run pricing variables (i.e., cash rebates and low financing offers). By so doing, we could explore the issues of the lag or lead effects between brands, and of the asymmetric share effects of the short-run price variables in the market.

Chi-square values were calculated according to the Pierce-Haugh Test statistics. The version used in this study was suggested by Ljung and Box (1978) and was also used for the study of Aaker, Carmen, and Jacobson (1982). Chi-square values which are significant at the 0.05 level are underlined, and those significant at 0.10 level are denoted by superscript b. Since we set the maximum lag at five, a significance chi-square means that a variable has a significant lag effect up to the five lag periods. The chi-square statistics are used to formally test the (asymmetric) lag effects of price promotions.

For the significance of each cross-correlation, a two-tail test was performed. Technically, the t-value for the cross-correlation which is significant at 0.05 level was underlined. Therefore, for example, the underlined values at lag 0 indicate that cross-correlations of two marketing variables are significant in the concurrent period. the underlined values at lead 2 mean that a price cut in the current period has a significant effect on the sales of two-periods-before. The test of asymmetric lead effects primarily employs the two tail test in this study, as used in Aaker, Carman, and Jacobson (1982) and Doyle and Saunders (1985).

5. Major Findings and Discussions

Overall, our results showed that the sub-compact car brands competed in a complicated fashion, showing the various dynamics of their short-run variables (especially cash rebate and low financing offer). Consistent with our prediction, however, a generalizable pattern regarding the direction of the lead and lag effects is found between high, medium, and low quality brands.

Major findings are reported in tables and discussed sequentially. An overall picture is presented first and then detailed discussions highlighting two cases follow as to the two major hypotheses.

5.1 Overall Results

The dynamic effects of price variables among major brands are summarized in Table 1.[5] The focus is given to looking at "woods" rather than "trees" by comparing how the two price promotion tools, rebate and financing, have the lead and lag effects between high, medium, and low quality brands.

Table 1. Summary of Overall Results

| | Affected by the prices of | | |
	High-quality brands	Medium-quality brands	Low-quality brands
LAG EFFECTS			
The shares of			
High-quality brands	2/8	0/12	0/16
Medium-quality brands	2/12	1/18	0/24
Low-quality brands	6/16	11/24	17/32
LEAD OR ANTICIPATION EFFECTS			
The shares of			
High-quality brands	1/8	3/12	1/16
Medium-quality brands	1/12	4/18	4/24
Low-quality brands	4/16	9/24	7/32

5. The details of cross-correlations and their t-values, and chi-square statistics are reported in the author's working paper at Concordia University (will be available upon request).

Table 1 shows how the shares of **high quality brands** are dynamically affected by the rebate and low financing of their own and the other medium and low quality brands. None of the chi-squares are shown to be significant for the effects of medium and low quality brands on the shares of high quality brands (0/12 and 0/16), suggesting that the shares of high quality brands were not significantly influenced by price cuts by medium and low quality brands. The shares of high quality brands are influenced by each other in the same category of high quality (i.e., 2/8).

An interesting pattern is identified when the above results for high-quality brands are compared with the results for the shares of medium and low quality brands, as Table 1 also shows how the shares of **medium quality brands** were influenced by the prices of high, medium, and low quality brands. Chi-square statistics show that the medium quality brands were not affected by any brand of low quality (0/24) while they were influenced by some brands of same (medium) quality and high quality (2/12 and 1/18).

The results for low-quality brands are consistent with the previous findings for high- and medium-quality brands. Table 1 clearly show that **low quality brands** were vulnerable to the price promotions (rebate or financing) by both higher (i.e., high and medium; 6/16 and 11/24) and the same (i.e., low; 17/32) quality brands.

The similar pattern is identified for *lead effects* as shown in the second part of Table 1. In summary, the overall picture of dynamic (both lag and lead) effects clearly indicates that asymmetry exists between high, medium, and low quality brands in the sub-compact car market, which is consistent with the existing studies. This study empirically shows that the directions of asymmetric (especially lag) effects favoured higher quality cars. This suggests that price promotions of higher quality brands in the current period significantly affected the shares of lower quality brands in both the current and subsequent periods, while the reverse was not found to be true in the car market.

It would be interesting to examine how the leading brand *Ford Escort*'s price promotion affected other high, medium, and low quality brands in the sub-compact car market. There was also a low quality brand, *Horizon* or *Omni*, which was aggressive in price promotion during the study period. The following discussions highlight the two cases to further explore the issue of asymmetric dynamic switching.

5.2 Lag and Lead Effects

Table 2 highlights how price promotions by an active medium-quality brand (*Ford Escort*) dynamically affected the shares of its own brand and the other brands of high, medium, and low quality. Cross-correlations between the two major price promotion variables of *Ford Escort*, R1 (rebate) and F1 (low financing), and the shares of the other brands in the car market (noted as s1, ..., s8, and s15) were reorganized in Table 2.

Table 2. The shares of other brands affected by price promotion of the leading brand (*Ford Escort:* a medium quality brand)

Share	Price Promotion Variables	Lead 3	2	1	Lag 0	1	2	3	Chi-square
High Quality Brands' Shares									
s2	rl	.34	-.04	-.05	-.02	.14	.03	-.09	1.36
	fl	-.02	-.18	-.06	-.37	-.00	.09	.07	4.98
s15	rl	.20	-.10	-.11	.18	-.02	-.11	.10	4.95
	fl	-.06	-.06	.01	.13	-.08	.07	-.01	10.36
Medium Quality Brands' Shares									
s1	rl	-.07	.32	.11	-.01	.04	-.03	-.11	1.99
	fl	.16	.16	-.04	.00	-.00	-.22	-.01	7.33
s6	rl	.08	.01	.19	.05	-.02	.03	-.04	0.66
	fl	.07	.28	-.08	-.17	.14	.05	-.30	11.11[b]
s7	rl	-.09	-.03	-.10	-.15	-.02	.01	.03	1.68
	fl	-.00	.01	.02	.05	.08	-.24	.21	8.05
Low Quality Brands' Shares									
s3	rl	-.00	-.05	.06	.00	-.13	.09	.11	3.57
	fl	.12	.05	.22	-.11	.02	.05	-.00	3.17
s4	rl	-.33	-.15	-.17	-.03	.02	-.09	.00	5.48
	fl	-.14	-.14	-.15	.57	-.13	-.00	-.02	25.27
s5	rl	-.34	-.19	-.16	-.00	.00	-.07	.04	5.07
	fl	-.12	-.15	-.09	.50	-.10	.05	-.07	20.40
s8	rl	.18	-.30	.19	.21	-.05	.03	.07	3.46
	fl	-.06	.10	.06	-.10	.00	.55	-.33	26.25

Chi-square statistics show that *Escort*'s short-run pricing efforts had significant lag effects on the shares of other brands, but not on the shares of any high quality-brands. *Ford Escort*'s financing offer had some significant lag effects on the shares of four same or lower quality brands (S4, S5, S6, and S8). The cross-correlations at lag 0 shows that S4 and S5 (the shares of Brand 4 and Brand 5), two domestic brands of lower quality, were significantly and negatively associated with *Escort*'s low financing offers (the cross-correlations with S4 and S5 at lag 0 are very high, i.e., 0.57 and 0.50).

Another important finding is that *Ford Escort*'s marketing efforts had significant *lead* effects on the shares of the other same or lower quality brands (see the underlined cross-correlations between R1 and s4, R1 and s5, F1 and s6, and R1 and s8). This shows that if people anticipated *Escort* to provide some incentives (rebates or low financing offers) in the near future, then the sales of the other brands (Brand 4, 5, 6, and 8) decreased in the current period because of the future deal anticipation.

Therefore the current analysis empirically highlights that a brand's price promotion or price cut through rebate or lower finance offers did have both lag and lead effects on the shares of the other brands in the market. These results seem to be consistent with the overall picture of the dynamic effects shown in Table 1, which shows that the *cross*-lead and lag effects, although seemingly complicated, exist among major brands in the market. Therefore, it is concluded that the first hypothesis is confirmed.

5.3 Asymmetric Share Effects

As shown above, *Ford Escort's* price promotion efforts had some significant *lead* effects on the shares of the other domestic brands. Table 2 showed the statistical significance of the underlined cross-correlations between R1 and s4, R1 and s5, F1 and s6, and R1 and s8, suggesting that the shares of Brand 4 and 5 in the current period are affected by the future rebate programs by *Ford Escort*. It is interesting to find that the brands vulnerable to *Escort's* incentive programs (i.e., Brand 4, 5, 6, and 8) are the brands of either lower quality or same shares. None of the higher quality brands were found to be negatively affected by *Escort's* announcement on the future deals.

Since the lead and lag effects were found to work in one direction in the *Escort's* case, i.e., from a higher quality brand to lower quality brands, we carefully examined the directions of lead and lag effects among all the brands in Table 1 so as to see if the asymmetric share effects could be generalizable in this market. Our analysis in this direction showed that the theory of asymmetric effects of price promotion appears to be generalizable in the sub-compact car market.

The issue of the asymmetric share effects is further examined by using the case of *Plymouth Horizon*, one of the low quality brands (according to *Consumer Report)*, which heavily employed the short-run incentive programs. Table 3 reports the cross-correlations between *Horizon's* short-run marketing efforts, R4 (rebates) and F4 (low financing) and the shares of other brands in the market. Although overall no lag and lead structures seem evident as in the study of Aaker, Carmen, and Jacobson (1982), some interesting patterns have been identified. First, the cross-correlations at lag 0 show that *Horizon* (a low quality brand) has significant effects only on some brands of equal (low) quality in the same period but did not have any significant effects on other higher quality brands. The finding strongly supports the idea of *asymmetric share effects* of the short-run price promotion such as rebates and low financing offers, which is consistent with the finding from the existing studies of *asymmetry*.

Chi-square statistics also show that the brand had a significant dynamic (lag) effect only on the other brands of the same low quality (s3, s4, and s5), not on the shares of higher-quality domestic brands. When it comes to the lead effects, the short-run incentives by *Horizon* had a few lead effects compared with *Escort's*. Again, it could not have a significant lead effect on the shares of higher-quality brands. Since this phenomenon was consistently found in the cases of the other brands including the

Escort's case discussed above, it is concluded that generally the price promotions of higher quality brands had the lead and lag effects on lower quality brands, not vice versa.

The above findings from a time-series analysis seem to support the idea of asymmetric share changes or asymmetric switching in the dynamic context. Specifically the results seem to support our main hypothesis 2, which states that a price promotion for higher quality brands has significant lead and lag effects on the shares of lower quality brands, but the reverse is not true. This current study empirically demonstrates that the asymmetric competition effects could exist even for durables such as cars as well as for frequently purchased non-durables such as grocery products.

Table 3. The shares of other brands affected by an active low quality brand, *Dodge Horizon*

Share	Price promotion Variables	Lead 3	2	1	0	Lag 1	2	3	Chi-square
High Quality Brands' Share									
s2	r4	.05	-.17	.13	-.28	-.13	.24	-.14	5.95
	f4	-.03	-.03	.00	-.18	.04	-.02	.14	2.29
s15	r4	-.13	.03	-.00	-.09	.05	.03	-.18	9.37
	f4	.09	-.03	-.11	.10	-.04	-.10	-.01	2.64
Medium Quality Brands' Shares									
s1	r4	.11	-.13	-.07	.05	-.31	.00	.02	8.74
	f4	-.02	.09	.03	-.10	-.08	-.02	-.14	3.34
s6	r4	.11	-.05	-.06	.17	-.16	-.22	-.12	9.99
	f4	.18	.16	.15	-.13	.09	.21	-.06	5.62
s7	r4	.12	-.24	.30	-.16	.01	.02	.14	4.18
	f4	-.00	.09	-.06	.12	.11	-.04	.00	3.38
Low Quality Brands' Shares									
s3	r4	-.07	-.00	-.05	-.02	.15	-.31	-.09	8.72
	f4	.23	.11	.21	.36	.05	.25	.14	13.93
s4	r4	-.13	.00	.04	-.03	.01	.06	.11	1.68
	f4	-.06	-.10	-.19	.39	.17	-.35	-.09	24.70
s5	r4	-.13	.00	.02	-.01	.03	.05	.13	1.75
	f4	-.01	-.13	-.15	.38	.11	-.32	-.08	21.00
s8	r4	-.12	.01	.08	-.05	.25	-.25	-.24	14.29
	f4	.19	.13	.13	.16	.07	.13	.10	4.15

5.4 Major Contributions and Implications

The test results of the asymmetric lead and lag effects could make the theory of asymmetric switching more generalizable. First, this paper empirically addresses asymmetric switching in the dynamic context. It shows that high-quality brands benefit from the asymmetric dynamic effects of their price promotions on the shares of lower-quality brands while low-quality brands do not affect the shares of higher-quality brands by lowering their prices. Second, this paper provides an empirical support for the asymmetric lead and lag effects for the case of a durable good (small cars), while the existing studies have focused mainly on the cases of frequently purchased grocery products. The two main results, combined together, could make the theory of asymmetric switching stronger and more generalizable. This effort will be important in enhancing marketing knowledge in that there have only been a few studies that have stressed the generalizations (Bass 1993; Leon and Schultz 1980; Weitz 1985; Zaltman et al. 1982, p.6).

The major result of this study, asymmetric lead and lag effects, also provides important managerial implications for practitioners (e.g., marketing managers). A manager of a low quality brand should not use price promotion (e.g., cash rebates or low financing offers) to take shares away from high quality brands. Consistent with the existing studies, the current study shows that a price reduction by a low quality brand will not take sales away from high quality brands, although it would steal sales from other low quality brands or lower quality brands.

A low-quality brand's price promotion could get rid of inventory problems, but should not be used as a competitive tool against high quality brands. However, a high quality brand could use price promotion as an effective competitive tool against lower quality brands. For example, I.B.M. could effectively use price promotion as a defensive tool against price attacks by low quality brands in the desktop computer market. AT&T could use it as a defensive strategy against low-quality competitors' attacks in terms of monetary savings in the long-distance call market.

6 Summary and Conclusions

Two major hypotheses were tested as part of this project. First, it was found that some significant *cross*-lead and lag effects of the short-run price variables on shares existed between different quality groups. In relation to the first hypothesis, it was also found that high quality brands were able to have the *cross*-lead and lag effects on the shares of other high quality brands and low-quality brands while the reverse was not true, supporting the theory of *asymmetric switching or share effects* of pricing variables using the time-series analysis. Although the asymmetric effects have been well documented for frequently purchased goods, no existing study has investigated the effects for durables.

It is not hard to find some real examples which show that asymmetric dynamic switching between brands exists also for the cases of non-frequently purchased products. We see that price promotions of high-quality brands in the current period have some significant effects on the sales of lower-quality brands both in the later periods and in the previous periods while the reverse is not true. In the non-frequently purchased product categories such as home furniture, men's suits, expensive stereo systems, etc, consumers wait until later when they expect higher-quality brands to be on sale in the near future. When faced by these anticipations, some of them switch from the purchases of lower-quality brands in the current period to the purchase of higher-quality brands in the near future, which produces the asymmetric dynamic effects (especially lead effects) between high- and low-quality brands. In most cases, the asymmetric switching including dynamic effects tend to work for higher-quality brands, and therefore against lower-quality brands. This current study empirically supports the asymmetric dynamic switching favouring higher-quality brands even for the non-frequently purchased products such as cars.

6.1 Limitations and Future Research Directions

Although the current study provides an empirical support for the *asymmetric cross*-lead and lag effects between brands, the theory of asymmetric effects needs to be further tested for other durables to be established as a strong theory. Because the current study looks at only a segment of the car market, we could test the theory using the data from the other segments of the car market (e.g., medium-size market). Further tests are needed for the cases of other durables, because consumer responses to price promotion for these durables would be quite different from those for the car market.

Another future direction would be to examine the asymmetric effects in relation to other important long-term variables such as distribution, pioneering, market share, etc. For example, asymmetric effects could exist between pioneer/first entrants and followers. The future studies could also examine the asymmetric effects of other short-run marketing variables such as advertising. An empirical study indicated that advertising cross-elasticities among major brands show analogous asymmetries (Carpenter et al. 1988). Economy brands are sensitive to their own advertising, but exert very little pressure on their national competitors. Therefore it would be interesting to further examine if short-run advertising has the asymmetric effects for other product categories in the same way as short-run price does. Furthermore, the interaction between short-run variables (e.g., advertising and price cut) and its asymmetric effect on sales or shares would be also interesting.

A future study could analyze the supply-side marketing interactions between brands by cross-correlating key marketing variables across brands, as Hanssens did (1980). The issue of asymmetry could be examined in the context of the supply-side marketing interactions or reactions. Low-quality brands' marketing would show more sensitive response to high-quality brands' marketing than high-quality brands' marketing to low-quality brands' marketing.

From the methodological point of view, another direction could be to analyze the endogeny of the short-run marketing variables, using the proposed time-series approach. Simply, the effects of brand shares on the marketing variables could be tested by the time-series cross-correlation analysis, using the Granger test (Granger 1969; Hanssens, Parsons, and Schultz 1990). This will make an important contribution since several estimation results from aggregate demand models including MCI (multiplicative competitive interaction) model and MNL (multinomial logit) model are expected to be biased if they treated some endogenous variables as exogenous.

This paper could not formally address why asymmetric switching occurs. An interesting direction would be to relate the asymmetric switching to consumer preference or taste and product positioning in a multi-attribute space. For example, DEFENDER model by Hauser and Shugan (1983) can be employed to address asymmetric switching or share changes since it explicitly incorporates the role of price in switching or share changes in the multi-attribute space.

As a conclusion, a time-series approach to short-run price competition and its asymmetric effects produced some important findings and appears to have a promising future. The theory of asymmetric switching supported in the present study must be tested in other markets to be stronger and more generalizable.

REFERENCES

Aaker, David, James Carman, and Robert Jacobson (1982), "Modeling Advertising-Sales Relationships Involving Feedback: A Time Series Analysis of Six Cereal Brands," *Journal of Marketing Research,* 19 (February), 116-25.

---------- (1991), *Managing Brand Equity,* The Free Press, New York.

Allenby, Greg (1989), "A Unified Approach to Identifying, Estimating, and Testing Demand Structures with Aggregate Scanner Data," *Marketing Science,* 8 (Summer), 265-80.

----------, Greg and Peter Rossi (1991), "Quality Perceptions and Asymmetric Switching Between Brands," *Marketing Science,* 10 (Summer), 185-204.

Automotive News (1982-87), Weekly, Crain Communications, Inc., Detroit, Michigan.

Bass, Frank and Thomas Pilon (1980), "A Stochastic Brand Choice Framework for Econometric Modeling of Time Series Market Share Behavior," *Journal of Marketing Research,* 17 (November), 486-97.

---------- (1993), "The Future of Research in Marketing: Marketing Science," *Journal of Marketing Research,* 30 (February), 1-6.

Blattberg, Robert and Kenneth Wisniewski (1989), "Priced-Induced Patterns of Competition," *Marketing Science,* 8 (Fall), 291-309.

Carpenter, Gregory, Lee Cooper, Dominique Hanssens, and David Midgley (1988), "Modeling Asymmetric Competition," *Marketing Science,* 7 (Fall), 393-412.

Day, George, Alan Shocker, and Rajendra Srivastava (1979), "Customer-Oriented Approaches to Identifying Product-Markets," *Journal of Marketing,* 43 (January), 8-19.

Doyle, Peter and John Saunders (1985), "The Lead Effect of Marketing Decisions," *Journal of Marketing Research,* 22 (January), 54-65.

Hanssens, Dominique (1980), "Market Response, Competitive Behavior, and Time Series Analysis," *Journal of Marketing Research,* 17 (November), 470-85.

----------, Leonard Parsons, and Randall Schultz (1990), *Market Response Models: Econometric and Time Series Analysis,* Boston, Kluwer Academic Publishers.

Jacobson, Robert and Franco Nicosia (1981), "Advertising and Public Policy: The Macroeconomic Effects of Advertising," *Journal of Marketing Research,* 18

(February), 29-38.

Judge, George, William Griffiths, Carter Hill, Helmut Lutkepohl, and Tsoung-Chao Lee (1984), *The Theory and Practice of Econometrics*, 2nd Edition, John Wiley and Sons, New York.

Kamakura, Wagner and Gary Russell (1989), "A Probabilistic Choice Model for Market Segmentation and Elasticity Structure," *Journal of Marketing Research*, 26 (November), 379-90.

Leone, Robert and Randall Schultz (1980), "A Study of Marketing Generalizations," *Journal of Marketing*, 44 (Winter), 101-18.

----------- (1983), "Modeling Sales-Advertising Relationships: An Integrated Time Series-Econometric Approach," *Journal of Marketing Research*, 20, August, 291-5.

Ljung, G.M. and Box, G (1978), "On a Measure of Lack of Fit in Time Series Models," *Biometrica*, 65 (Fall), 297-303.

Maddala, G. (1978), *Econometrics*, McGraw-Hill Book Company, London.

Mills, Terence (1990), *Time Series Techniques For Economists,* Cambridge University Press, Cambridge.

Russell, Gary and Ruth Bolton (1988), "Implications of Market Structure for Elasticity Structure," *Journal of Marketing Research,* 25 (August), 229-41.

Urban, Glen, Philip Johnson, and John Hauser (1984), "Testing Competitive Market Structures," *Marketing Science*, 3 (Summer), 83-112.

Weitz, Barton (1985), "Introduction to Special Issue on Competition in Marketing," *Journal of Marketing Research,* 22 (August), 229-36.

Zaltman, Gerald, Karen Lemasters, and Michael Heffring (1982), *Theory Construction In Marketing: Some Thoughts on Thinking,* Theories in Marketing Series, John Wiley and Sons, New York.

Profit Impacts of Aggressive and Cooperative Pricing Strategies

Martin Natter[1] and Harald Hruschka[2]

[1] Vienna University of Economics, Austria
[2] University of Regensburg, Germany

Abstract. We study profit impacts of aggressive/cooperative pricing strategies in a dynamic oligopolistic environment. A logistic model with asymmetric reference prices is used to forecast market shares. Pricing strategies - optimized by simulated annealing - are evaluated by simulating the empirical price distribution of competitors. It is shown that there are regions on the aggressiveness/cooperation path that a rationally operating manager would prefer to others, namely where his position is strongest as compared to the position of all rivals.

Keywords. Pricing research, competitive strategy, aggressive strategies, asymmetric reference prices

1 Introduction

Competitive dynamic models have been frequently used in the recent literature [e.g. Villas-Boas (1994)] to investigate optimal marketing strategies. Most papers focus on optimum (product line) pricing or advertising strategies in a competitive environment. Often found model assumptions are Nash equilibria and Stackelberg leader-follower hypotheses. An open question is whether for frequently purchased consumer goods assumptions like optimal reactions of competitors can be found as often in real world situations as one would expect due to the growing literature in this field. On the other hand, one can expect reactions from a firm if its profits are significantly affected by the permanent aggressive strategy adapted by another company. To assess the probability of competitive reactions one should analyze possible consequences of different strategies.

Though the price optimization performed by a producer has side-effects on the profits of his competitors, this is typically not modeled in the objective function. Our intention is not only to study the side-effects but also to investigate possibilities of influence. Deviations from optimum product line prices could be accepted if, for instance, a more aggressive strategy helps to improve other management goals like dominating a certain segment by ousting a rival product from that segment.

If one possible instrument to oust a rival product from a segment is an aggressive pricing strategy managers are, however, faced with certain difficulties: they have to decide about the extent of possible aggressive business practices and repercussions on their own profits. It is shown that a firm may be much better off if it weights the degree of aggressiveness within a certain range, because a strategy might become very costly when it exceeds that range. To avoid reactions of rivals it may also be useful to investigate which policies will have moderate or strong profit impacts on all products.

2 Models and Analysis

2.1 A Logistic Model with Asymmetric Competitive Reference Prices

The reference price theory [Tolman (1951), Muth (1961)] provides a behavioristic explanation of a price variation effect that is not caused by the absolute price only. A reference price [Tull, Boring, and Gonsior (1964)] is an internal price which the consumer compares to the prices observed [Winer (1988)]. It is assumed that the consumer has in mind price information gathered on previous purchase occasions (hence 'internal' price) and uses it by way of comparison when making new purchase decisions. If the current price is below the reference price, then it is judged favorably. If, on the other hand, the current price is higher, it is judged unfavorably. For an overview of the relevant literature, see e.g. Winer (1988) or Bridges, Yim, and Briesch (1993).

While studies of price competition model competitive prices most empirical studies of reference price models only consider deviations of the observed price from the reference price of a product. If competitive prices influence sales of a product it is to be expected that also deviations of observed competitive prices from their reference price have an impact on sales. Therefore, the model considered here contains hypotheses such as significant effects of relative prices $(p_{i,t})$, customer holdover effects $(S_{i,(t-1)})$ as well as asymmetric deviations from a brand's reference price and competitive brands' reference prices. The market share of brand i in period t is given by the following relation:

$$S_{i,t} = f(\xi_{i,0} + \xi_{i,1} S_{i,(t-1)} + \sum_{j=1}^{J}(\xi_{i,(1+j)} p_{j,t} + \xi_{i,(1+J+j)}(p_{j,t}^r - p_{j,t})^-$$

$$+\xi_{i,(1+2J+j)}(p_{j,t}^r - p_{j,t})^+)) \tag{1}$$

where f() denotes the logistic function. Reference prices p^r are a function of a constant, the previous relative price of a brand and a trend term:

$$p^r_{i,t} = \theta_{i,0} + \theta_{i,1} p_{i,(t-1)} + \theta_{i,2} t \tag{2}$$

ξ and θ denote parameters of the market share equations and the reference price equations, respectively. Predictors as well as dependent variables are linearly transformed into the interval $[0; 1]$.

If the trend coefficient is positive, then in future periods higher price expectations are formed than in past periods. When a strategy of constant prices is assumed, increasing deviations from the reference price occur. In the context observed, i.e. that negative deviations of the reference price from the current price have a positive effect on the own sales volume and a negative effect on the sales of rival products, with time (ceteris paribus) increasing market shares are to be expected. In case the deviations from the reference prices adversely affect the profit, with time the price would have to be increased (proportionally to the effect of the trend coefficient) to meet expectations. A negative trend coefficient is worthy of analogous considerations, the prices having to be decreased with time. Parameters $\xi_{i,(1+j)}$ with $(i = j)$ measure the price effects of product i and have a negative sign as high prices have a negative impact on the market share. Competitive price coefficients $\xi_{i,(1+j)}$ with $(i \neq j)$ have a positive sign in the case of substitutional relations and a negative sign in the case of complementary relations. If the own price is higher than expected by the consumers (this refers to parameters $\xi_{i,(1+J+j)}$ with $(i = j)$), we anticipate a negative impact on the sales volume of product i. If the own price is lower than expected (measured by parameters $\xi_{i,(1+2J+j)}$ with $(i = j)$), then a sales promoting effect is produced. For the competing brands $(i \neq j)$ it is exactly the other way round. High coefficients measuring the price differences make us anticipate a pulsation strategy always endeavouring to keep the prices below the demanders' expectations.

Data

For the empirical study of the models a scanner database consisting of time series with 73 weekly observations each is available. The times series comprise sales values and prices collected in 4 retail stores of one chain and referring to 7 brands of a product category (consumer nondurable).

Because of the existing competitive relationships the specified models contain a rather high number of parameters, and we pool the outlets for estimation purposes. One observation per outlet is lost because of modelling reference prices and carry-over effects. Therefore 288 observations are available for each product. A higher number of degrees of freedom results in higher coefficient reliability. Another advantage of pooling is said to be a reduced risk of multicollinearity due to the variability that usually is higher in cross-sectional data than in longitudinal-sectional data.

Empirical Results

For almost each product it is possible to observe influences due to deviations between the prices observed and the expected price notions of the consumers. Modelling competitive prices and their deviations from the prices expected by the demanders shows that although normally a product does not compete with all other products, there are a few potential rivals for the consumers to choose from.

The logistic form of the model has an average variance explanation of $R^2 = 67.8\%$ (for the individual R^2, see table 2). Models are selected by minimizing the Prediction Criterion (PC) which relates the number of parameters and variance explanation:

$$PC = \frac{SSE}{N-P}(1 + \frac{P}{N}) \qquad (3)$$

N symbolizes the number of observations, P the number of parameters and SSE the sum of squared errors.

After a stepwise backward selection we obtain a model with less than 5 price/promotion parameters per equation. The results strongly emphasize the relevance of the reference price concept. Customer holdover effects are insignificant for all equations.

The reference price model is a good choice between the number of parameters and the predictive power when inserted in the market response model. It was selected from more encompassing models that considered longer influences in time (ref-model 1) as well as other predictors (ref-model 2). Table 1 gives a summary of the reference price models considered, where Q_t denotes the volume of an outlet at time t.

Table 1: Price expectations

Model	Equations
ref-model 1	$p^r_{i,t} = \theta_{i,0} + \sum_{\tau=1}^{5} \theta_{i,\tau} p_{i,(t-\tau)}$
ref-model 2	$p^r_{i,t} = b_{i,0} + \sum_j b_{i,j} p_{j,t-1} + b_{i,J+1} S_{i,t-1} + b_{i,J+2} Q_{t-1} + b_{i,J+3} t$
final model	$p^r_{i,t} = \theta_{i,0} + \theta_{i,1} p_{i,(t-1)} + \theta_{i,2} t$

Table 2: Variance explanation of the reduced market share model

Product	R^2	avg.share
1	0.808	26.8
2	0.531	7.1
3	0.718	16.4
4	0.754	12.3
5	0.772	16.7
6	0.548	5.8
7	0.615	14.8

2.2 Measurement of Reactions

General statements about an optimal price in a competitive situation cannot be made. In an oligopolistic market the optimal price depends on the assumptions made about the competitors' behavior and is, therefore, also suited for reaction function approaches or game-theoretic studies [for an overview see Vilcassim, Kadiyali and Chintagunta (1995), Corfman and Gupta (1993) or Moorthy (1993)]. Hence, it is to be investigated whether realistic assumptions do exist. It must thus be discussed which assumptions could be appropriate for the market under study.

Here it would be interesting to know whether the firms observe the pricing policy of their rivals and react to it. If this is the case and if price changes made by a firm systematically lead to reactions that can, therefore, be empirically proved, these reactions would have to be included into the price optimization. For this purpose, we will examine with the aid of our data whether such a behavior can be found in the market under review:

$$(P_{i,t} - P_{i,(t-1)}) = f\left(\theta_{i,0} + \sum_{j \neq i}\sum_{\tau=2}^{T} \theta_{i,j,\tau}(P_{j,(t-1)} - P_{j,(t-\tau)})\right) \qquad (4)$$

Equation 4 considers the price change of a product as compared with the preceding period as a function of lagged price changes of rival products. $f()$ represents both, linear and logistic functions. If, when estimating this equation, significant values are obtained for parameters $\theta_{i,j,\tau}$, that indicates that price changes of product i trigger (lagged) reactions of the competition. When optimizing the pricing strategy, these significant parameters can be used to control the competitors' behavior.

Empirical Findings

For the empirical test of equation 4, the maximum lag (T) was assumed to be 6 periods. The estimation of the model under consideration does not provide significant parameters so that the reaction function approach is not confirmed. That may well be due to the fact that retail chains are given plenty of scope when fixing the prices and that there are no (lagged) responses to their own actions. Another explanation may be that there are reactions rather in the long term than in the short term as stated here.

For lack of more detailed information, for the optimum pricing we assume that the competitors (during the planning horizon) sell their products at the mean price observed. Taking account of the empirical distribution of deviations from average relative prices of all products the assumption of constant competitive prices seems unrealistic and the evaluation of the strategies at average competitive prices could lead to unrealistic results. Therefore, we finally evaluate the strategies at competitive prices drawn from the empirical distribution of the observed prices.

2.3 Optimization Model

As in this market producers typically sell 2 or more brands of a product category, product line pricing must be performed. As the model that tests price reactions to pricing activities of other firms failed to capture any significant effects, we assume average competitive prices. We plan strategies for a duration of $T' = 30$ discrete periods.

If a manager decides to act more forcefully on the market, the degree of aggressiveness should be optimal in terms of difference between losses regarding every competitive product and the firms costs (measured in deviations from an optimal strategy). The aggressive strategy that leads to the strongest position of firm k with produkts i as compared to the rival firms l with products j results from the maximization of equation 5:

$$\sum_j \left(\left(\pi_{l(\zeta_{l,i}=1,\zeta_{l,j}=0)} - \pi_{l(\zeta_{l,i}=1,\zeta_{l,j}\neq0)} \right) - \left(\pi_{k(\zeta_{k,i}=1,\zeta_{k,j}=0)} - \pi_{k(\zeta_{k,i}=1,\zeta_{k,j}\neq0)} \right) \right) \quad (5)$$

To guarantee a stronger position against all other competitors and to avoid that a rival gets better results than the company itself, this difference should not only be calculated for the sum of losses of products of competitors but for each product separately.

We assume that every firm k maximizes its profit as follows:

$$\pi_k = \sum_j \zeta_{k,j} \sum_{t=1}^{T'} \left(\left(Q_{j,t}(P_{j,t} - C'_j) \right) (1+r)^{-t} \right) \tag{6}$$

where C'_j denotes constant marginal costs (i.e. a linear cost function) of brand j, and $Q_{j,t}$ the sales for brand j in period t for a given price strategy. In this approach, we use $\zeta_{k,j}$ to weight the profits of competitors. For the brands of the producer under consideration, we use $\zeta_{k,i} = 1$. To study different degrees of aggressiveness, we vary $\zeta_{k,j}$ between $\zeta_{k,j} = -1$ (one unit less profit for a competitor equals one additional unit of own profits) and $\zeta_{k,j} = +1$ (cooperative strategy; one additional unit of competitors' profits equals one additional unit of own profits). The prices are restricted by the minimum and maximum observed values. The capitalization rate r depends on the best alternative investment and is assumed to be $r = 0.1\%$ per period (week).

One difficulty is that usually there is no information available on the costs competitors have to meet. On the assumption that producers sell their products, on average, for an optimum price (for the simple linear statical case), we use the Amoroso-Robinson relation to obtain an estimation of the variable costs of competitors. The estimate of such variable costs may be improved by adjusting them by the fraction of the known variable costs of the own products and the variable costs estimated on this assumption.

In the following, we discuss the impacts of different ζ- levels on all profits for our model validated on scanner data. For the price optimization of the dynamic nonlinear market response functions, we use Simulated Annealing [see Kirkpatrick et al. (1983)]. Simulated Annealing is an iterative method that accepts improvements of the target function $((\pi^{new} - \pi^{old}) > 0)$ with probability 1. To escape from local minima a stochastic component (see equation 7) is introduced which permits - with deminishing probability - the acceptance of worse solutions than the present one. Starting with a c-value that allows to reach any possible state in the solution space, the following steps are repeated until the target function stabilizes:

1. randomly select a decision variable (period/product)

2. generate a new random value for this variable (calculate new ref. prices)

3. evaluate the new strategy

4. accept the new strategy if $(\pi^{new} - \pi^{old}) > 0$ or

$$(1 + e^{-(\frac{(\pi^{new} - \pi^{old}}{c})})^{-1} > U[0; 1] \tag{7}$$

5. reduce control parameter c

$U[0; 1]$ denotes uniformly distributed random numbers between zero and one.

2.4 Aggressive/Cooperative Product Line Pricing

In a first step, a producer has to find out which of the competing brands his brands are really in competition with. One possibility to discover where competition is really going on may be to study the matrix of cross-price elasticities. If there is a complex structure of interactions with many significant parameters, the matrix of cross-price elasticities can be a little confusing, and a graph showing the interactions may be more helpful. Figure 1 shows

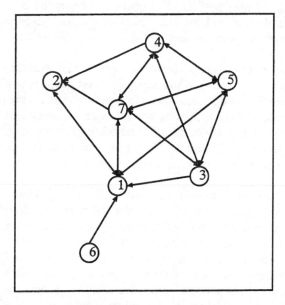

Figure 1: Relations between products: result of a MDS based on the cross-price-elasticities of the logistic static competitive model

the products in 2 dimensions in a way that products with high cross-price elasticities[1] are close to each other[2]. A product depends on the policy of those products to which arrows are pointing. The higher the number of arrows starting from a product and the closer its position to other products the more it can be influenced by the others. On the other side, products where

---

[1] Calculated at mean prices for the static logistic competitive model which is not further discussed here.

[2] Distances between two products i and j are calculated by $(\Gamma - 0.5 * (\epsilon_{ij} + \epsilon_{ji}))$, with $\Gamma > \max(\epsilon_{ij})$. Γ is used for rescaling. As the relations are typically asymmetric, we calculate the arithmetic mean of ϵ_{ij} and ϵ_{ji}.

a lot of arrows from other products point to (like product 1) have a bigger potential to influence the profits of the other products.

The 7 brands considered here dominate the market. They are made by 3 firms ($k_1 = \{1, 2\}$, $k_2 = \{3, 5, 7\}$ and $k_3 = \{4, 6\}$). So each firm k is in a position to manipulate the prices of 2 or 3 brands of the product category (detergents) considered.

In this study, we investigate profit impacts for a scenario where k_2 optimizes pricing strategies of its product line (3, 5, 7). Figure 1 shows that the biggest rivals of k_2 are products 1 and 4 which are close to (3, 5, 7) and interact with them. Therefore, profits of products 1 and 4 are of high interest for firm k_2 and should be considered in the target function (6). However, it is not clear to which extent k_2 should try to negatively affect its rivals, if it should do so at all. To find out about that we vary zetas between -1 and +1 and have to decide about the extent based on the information provided by a variation of $\zeta_{2,j}$. As every influence involves costs, real rivals should be selected carefully. Therefore, profits of products 2 and 6 are not considered in the target function (i.e. $\zeta_{2,2} = 0$ and $\zeta_{2,6} = 0$), allthough they may be influenced by (side-)effects of the optimization.

Figure 2 shows the changes of profits for k_2 and products 1 and 4 as compared to the profits when ignoring the rival profits. All results reported below are mean values over 5000 different random competitive strategies. Allthough, our focus is on profit impacts and not the specific strategies, we want to summarize some characteristics found. More aggressive strategies are typically characterized by:

- lower mean prices than cooperative strategies

- stronger price reductions if that negatively affects the rival products

- smaller price increases where rival products gain from price increases.

If a rival has a strong market position (like product 1) it is difficult and costly to be aggressive against him. In this market it is also expensive to cooperate with product 1 (high relative prices and high market share) because in case of cooperation (higher average prices) many customers switch to product 1. The optimal degree of aggressiveness as proposed by equation 5 is at $\zeta_{2,1} = -0.1$ and $\zeta_{2,4} = -0.3$. Figure 3 shows these differences from optimum profits of k_2 for negative zetas for products 1 and 4.

At about $\zeta_{2,4} = -0.76$ the difference between $\pi_{3,5,7}$ and π_4 is zero. More aggressive policies (i.e. $\zeta_4 < -0.76$) would weaken the relative position of firm k_2. The profit relation against product 1 as compared to an indifferent

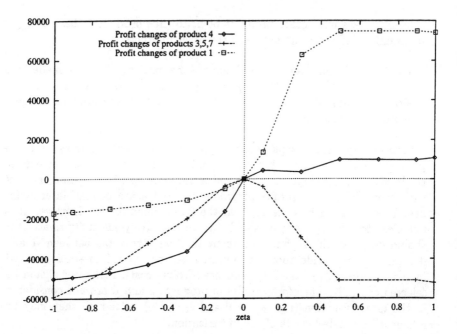

Figure 2: Differences in profits of firm k_2 and products 1 and 4.

policy ($\zeta_{2,1} = 0$) would already get worse at about $\zeta_{2,1} = -0.13$. The results show that firm k_2 should be less aggressive against product 1 (avg. sales share is 26.8%) than against product 4 (avg. sales share is 12.3%). If the average price[3] of a brand is taken as a measure for the quality of a brand, this result would support the statments of Blattberg et al. (1995), i.e. that higher quality brands can gain a competitive advantage against lower quality brands by aggressive promotion policies. Even by using the policies of all its brands (3, 5, and 7) firm k_2 can hardly improve its position against brand 1.

If one firm is less powerful than another and the reaction of this rival can be dangerous, it may prefer a cautious policy avoiding negative consequences for the rival.

---

[3]The average relative prices of brands 1 and 4 are $\bar{p}_1 = 1.11$, and $\bar{p}_4 = 1.04$, respectively. The average prices of brands 3, 5, and 7 are 0.99, 1.07, and 0.73.

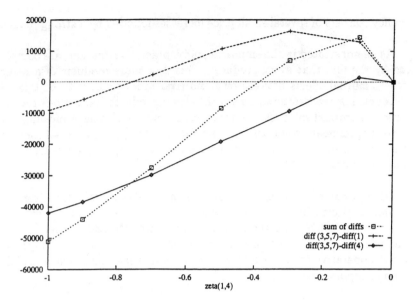

Figure 3: Relative position of firm k_2 as compared to differences for product profits of 1 and 4

3   Conclusion

A reference price model with asymmetric reference and competitive (7 products) reference prices was validated on scanner data (avg. R^2=67.8%). Every product has one or more significant reference prices.

A model we estimated to test competitive price reactions showed no significant effects. Therefore, average prices and no price reactions were considered for the optimization problem. The evaluation of the optimized strategies at competitive prices generated from the empirical distribution of prices shows that a slight degree of aggressiveness results in the strongest position of firm k_2 against product 1 which has a strong position in that market. A slightly more aggressive policy is optimal against product 4.

If sales of product j significantly depend on positive deviations from competitive reference prices (i.e. $(p_i^r - p_i) > 0$) a strategy that only slightly increases prices of i (in case of pulsing) reduces the benefits for the rival (j) products. On the other hand, if there are significant effects of negative deviations from competitive reference prices (i.e. $(p_i^r - p_i) < 0$) an aggressive policy of i would always try to be below its reference prices. So, in case of significant asymmetric competitive reference price influences, an aggressive policy consists of several small price increases followed by a strong reduction.

As demonstrated in the empirical study, practical situations are complicated by the fact that every producer offers 2 or more products of a category where mutual effects between own and rival products have to be taken into account. For realistic situations with several products and periods the calculation of optimal solutions is unpracticable and Simulating Annealing, which gives approximate solutions can be used. While approximate solutions may be second best solutions they may have the advantage to be less transparent for customers.

When a company wishes to influence the profits of its competitors or to favour a specific brand of its product line (because of its long-term product policy), calculating the influences of aggressive and cooperative pricing policies (or advertising strategies) and deciding on which aggressiveness/cooperative (zeta) level it wants to operate may represent a useful decision support.

References

Blattberg, R. C., R. Briesch, and E. J. Fox, 1995, "How Promotions Work", *Marketing Science*, Vol. 14, No.3, 2/2, 122–132.

Bridges, E., C. K. Yim, and R. A. Briesch, 1993, "A High-Tech Product Market Share Model with Customer Expectations", Working Paper Nr. 93, Jesse H. Jones Graduate School of Administration, Rice Univ., Houston, TX 77251.

Corfman, K.P. and S. Gupta, 1993, "Mathematical Models of Group Choice and Negotiations", in Nemhauser, G. L., A. H. G. Rinnooy Kan, J. Eliashberg, G. L. Lilien, (Eds.): *Handbooks in Operations Research and Management Science: Marketing*, Vol. 5, Amsterdam, North-Holland.

Kirkpatrick, S., C. D. Gelatt, and M. P. Vecchi, 1983, "Optimization by Simulated Annealing". *Science*, Vol. 220, No. 4598, 671–680.

Moorthy, K. S., 1993, "Competitive Marketing Strategies: Game-Theoretic Models", in Nemhauser, G. L., A. H. G. Rinnooy Kan, J. Eliashberg, G. L. Lilien, (Eds.): *Handbooks in Operations Research and Management Science: Marketing*, Vol. 5, Amsterdam, North-Holland.

Muth, J. F., 1961, "Rational Expectations and the Theory of Price Movement", *Econometrica*, 29, 315–335.

Rao, V., 1993, "Pricing Models in Marketing", in Nemhauser, G. L., A. H. G. Rinnooy Kan, J. Eliashberg, G. L. Lilien, (Eds.): *Handbooks in Operations Research and Management Science: Marketing*, Vol. 5, Amsterdam, North-Holland.

Tolman, E. C., 1951, *Behavior and Psychological Man*. Berkeley, CA, Univ. of Calif. Press.

Tull, D. S., R. A. Boring, and M. H. Gonsior, 1964, "A Note on the Relationship of Price and Imputed Quality", *Journal of Business*, 38 (April), 180–191.

Villas-Boas, J. M., 1993, "Predicting Advertising Pulsing Policies in an Oligopoly: A Model and Empirical Test", *Marketing Science*, 12, 88–102.

Vilcassim, N. J., V. Kadiyali, and P. K. Chintagunta, 1995, "Investigating Dynamic Multifirm Market Interactions in Price and Advertising: A Conjectural Variations Approach." Preprint volume of the *International Workshop on Dynamic Competitive Analysis in Marketing*, September 1-2, Montreal, Canada.

Winer, R., 1988, "Behavioral Perspective on Pricing: Buyers' Subjective Perceptions of Price Revisited", Devinney, T. (Ed.), *Issues in Pricing*. Lexington Books, 35–57.

Strategic consumers in a durable-goods monopoly

Jacques Thépot

Université Louis Pasteur, Strasbourg, France

Abstract

This paper is aimed at exploring the dynamic game structure inherent to the durable-goods monopoly and the Coase conjecture. The problem is stated as a two-level dynamic game between the monopolist and n identical consumers embedded in a given aggregate demand. A two-period model close to Bulow's model suggests that the monopsony power of the consumers could help the seller partly to avoid the intertemporal price discrimination ; the continuous time problem is analysed in a two-level differential game framework. In this context, a direct proof of the Coase conjecture is given.

1. Introduction

The Coase conjecture (1972) postulates that a durable-goods monopolist can lose his monopoly power because the consumers are expected a fall of the price in the future and are unwilling to pay more than the competitive price for the early units of the good. A key ingredient of this phenomenon is the length,Δ, of real time between the periods within which the firm is committed not to change its price. Coase argues that the monopoly intertemporal profit vanishes when $\Delta \to 0$. Significant works have been devoted in recent years to this theme : Bulow (1982) develops a two-period model where it is proved that the monopolist is better off when leasing when selling the good : this illustrates the role of the commitment period indicated above. The contribution of Stokey (1981) provides an in-depth analysis of the problem : Assuming that , at each date, the buyers form expectations about the total stock of the good that will have been sold by each date in the future, this author defines various types of equilibria in terms of buyers' expectations along which the Coase conjecture is satisfied. Gul, Sonnenschein &

Wilson (1986) get more general results in a price-setting context where the monopolist faces a continuum of consumers. Sobel (1991) examines the influence of the entry of new consumers on the market, Van Cayseele (1991) introduces consumer rationing rules. This paper is aimed at exploring in a systematic way the game structure inherent to the durable-goods problem : clearly, this problem can be seen as a *dynamic game* between the monopolist and the consumers and all the above mentioned works are strongly related to this point of view, sometimes implicitly. Of course, the key difficulty is to formalize the behavior and the expectations of the consumers, to see how they interact through time with those built by the producer and particularly to determine the demand for the durable. An additional difficulty arises from the continuous time formulation which is requested to cope with the core of the Coase conjecture. Our starting point is to consider a two-level dynamic game between a durable-goods monopolist and a set of n identical consumers. Contrary to the conventional literature on durable-goods, any consumer is a *multi-unit buyer* ; this situation deals with various industrial long-term customer-supplier relationships commonly implemented in business practice for equipment goods (e.g. mainframe computers). In this context, the consumers are allowed to manage the intertemporal strategic interactions with the producer ; this way, they become *strategic consumers*. In this two-level dynamic game, the producer acts as the leader who fixes at any date the price of the durable, while the consumers act as followers who choose the quantities. As it is known in dynamic game theory (Basar & Olsder, 1982), two types of equilibria can be defined :

- The open loop equilibrium, where the players are committed to strategies determined ex ante,

- The closed loop equilibrium, where the strategies are time-consistent.

Hence the paper is organized as follows : a two-period model analogous to Bulow's one (with utility functions leading to a given aggregate linear demand) is developed in section 2. The main results are the following : (i) the open loop strategy of the monopolist consists of selling the good only in the first period ; it gives the same profit as the leasing strategy does. (ii) in closed loop, the monopolist sells on both periods, but *the profit is higher than in open loop for small values of the number of consumers, n.*(iii) when $n \to \infty$, the closed loop solution converges to the solution given by Bulow. Such surprising results contradict the "leasing is better than selling" argument supporting the Coase conjecture. This phenomenon is easily interpreted in our game context : in closed loop, the consumers are allowed to consider that postponing a purchase order in the first period will decrease the stock of durable held at the beginning of the second period and, consequently, will increase the price charged by the producer in this period. This behavior is related to the *monopsony power* of the consumers which makes them aware of the impact of their purchase decisions on the future supply ; clearly such a power ought to decrease when the number of consumers on the market increases[1]. The analysis is then extended in section 3, with an infinite horizon continuous time model using *two-level differential games* techniques. The utility function of any identical consumer is only assumed to satisfy standard conditions of regularity and concavity and to lead to an aggregate demand of given size. In this general setting, it is proved that the open loop equilibrium has the same features as in the two-period model while the Coase conjecture is true for the closed loop equilibrium : this suggest that the monopsony power cannot offset the intertemporal price discrimination in a continuous time situation ; incidentally, the Coase conjecture is established without specific assumptions on the consumers expectations.

2. A two-period example

Let us consider an unique firm selling a durable good on two successive periods to a set of n customers. Let q_t^i the quantity bought at period t by consumer i, for $i = 1, ..., n$ and $t = 1, ..., T$, with $T = 2$. Customer i's surplus, S^i, is defined by a intertemporal function of the form :

$$S^i = u(q_1^i) - p_1 q_1^i + \delta \left(u \left(q_1^i + q_2^i \right) - p_2 q_2^i \right), \tag{2.1}$$

where δ denotes the discounting factor and u the one-period utility function, which are assumed to be identical for all the consumers. Since there is neither depreciation nor technological obsolescence of the durable good, utility enjoyed at period t depends on the cumulated quantity, $Q_t^i = \sum_{s=1}^{t} q_s^i$, purchased by consumer i to date. Utility function u is assumed to be increasing and strictly concave

---

[1] A converging analysis in the cases where the monopolist faces a finite number of one-unit buyers can be found in Bagnoli &al.(1989). Independent works by Dudey (1993) on a similar linear-quadratic model provide very close results.

and such that function $Q \to Qu'(Q)$ is strictly concave. In the following, we shall restrict the analysis to the linear-quadratic case, where the utility function is given by (for $Q \le 1/n$) :

$$u(Q) = Q - nQ^2/2, \tag{2.2}$$

so as to get a similar model than Tirole (1988, p.81). The aggregate "rental demand function", $D = nu'^{-1}$, does not depend on n. We assume that there are no costs of production, so that the profit function of the firm is :

$$\Pi = p_1 q_1 + \delta p_2 q_2. \tag{2.3}$$

where $q_t = \sum_{i=1}^{n} q_t^i$ stands for the quantity sold at period t. The problem is stated as a leader-follower game where the prices are determined by the producer and the quantities purchased at period 1 and 2 are chosen by the consumers. Two types of equilibrium can be considered (cf. Basar & Olsder, 1982):

The open loop equilibrium where the strategy of the producer is defined by a vector of prices (p_1, p_2) charged at period 1 and 2 ; the strategy of consumer i is a follower strategy defined as a *vector* of quantities, function of the prices, i.e. $q_t^i = q_t^i(p_1, p_2), t = 1, 2.$:

The closed loop equilibrium where the dynamic structure of the game is fully considered : each player is allowed to observe the state of the system at any stage of the dynamic game where he has to take a decision :

- in period 1, the producer fixes price p_1 and consumer i chooses the quantity to be purchased knowing this price, i.e. $q_1^i = q_1^i(p_1)$,

- in period 2, the strategy of the producer is determined as a function of the cumulated sales. The price charged in period 2 is defined through a **feedback rule**, $p_2 = p_2(q_1^1, ..., q_1^n)$ and each customer then fixes his price $q_2^i = q_2^i(q_1^1, ..., q_1^n, p_2)$.

2.1. The open loop equilibrium

The strategy of consumer i results from determining the optimal response $\{q_1^i, q_2^i\}$ to any price program $\{p_1, p_2\}$, namely :

$$\max_{q_1^i, q_2^i} \left[\left(u\left(q_1^i\right) - p_1 q_1^i \right) + \delta \left(u\left(q_1^i + q_2^i\right) - p_2 q_2^i \right) \right] \tag{2.4}$$

Optimality conditions of program (2.4) are :

$$u'\left(q_1^i\right) - p_1 + \delta u'\left(q_1^i + q_2^i\right) = 0. \tag{2.5}$$

$$u'\left(q_1^i + q_2^i\right) - p_2 = 0, \tag{2.6}$$

which yields :

$$p_1 = u'\left(q_1^i\right) + \delta p_2. \tag{2.7}$$

Clearly $u'(Q)$ measures **the willingness to pay** of any consumer to enjoy Q units of the good over one period ; Relation $r = u'(Q)$ defines the rental demand (in the inverse form) used in durable-goods literature (cf. Stokey, 1981) : in any period, the consumption level of any consumer is such that the current price is equal to the discounted sum of the rental prices over the future The strategy of the producer has to incorporate the best response functions of the n consumers over periods 1 and 2, defined by relations (2.5) and (2.6). It amounts to solving program (2.8), where quantities q_1^i and q_2^i are formally considered as "decision" variables of the producer :

$$
\begin{aligned}
&\max_{p_1,p_2,q_1^i,q_2^i} \left(\sum_{i=1}^{n}\left(p_1 q_1^i + \delta p_2 q_2^i\right)\right) \\
&s.t. \\
&u'\left(q_1^i\right) - p_1 + \delta u'\left(q_1^i + q_2^i\right) = 0, \\
&u'\left(q_1^i + q_2^i\right) - p_2 = 0, \\
&i = 1, ..., n.
\end{aligned}
\tag{2.8}
$$

Since the consumers are identical, at the optimum of program (2.8), $q_t^i = q_t/n$ and necessary conditions lead to the following relations :

$$
\begin{aligned}
&p_1 + (q_1/n)\left(u''\left(q_1/n\right) + \delta u''\left((q_1 + q_2)/n\right)\right) \\
&+\delta\left(q_2/n\right)u''\left((q_1 + q_2)/n\right) = 0, \\
&p_2 + (q_1/n)\,u''\left((q_1 + q_2)/n\right) \\
&+ (q_2/n)\,u''\left((q_1 + q_2)/n\right) = 0.
\end{aligned}
\tag{2.9}
$$

Relations (2.5),(2.6) and (2.9) yield :

$$u'\left(q_1/n\right) + (q_1/n)\,u''\left(q_1/n\right) = 0, \tag{2.10}$$

$$u'\left((q_1 + q_2)/n\right) + \left((q_1 + q_2)/n\right)u''\left((q_1 + q_2)/n\right) = 0. \tag{2.11}$$

Relation (2.10) indicates that $q_1^i = q_1^M$, where q_1^M is the one-period bilateral monopoly[2] quantity. Since function $Qu'(Q)$ is strictly concave, relation (2.11) implies $q_2^i = 0$. We have $p_1 = (1 + \delta)u'(q_1^M)$, $p_2 = u'(q_1^M)$. In the linear quadratic case, we have $q_1^M = 1/2n$ and the **equilibrium strategies** are $:p_1 = (1 + \delta)/2$, $p_2 = 1/2$, $q_1 = (1 - p_1 + \delta p_2)$, $q_2 = p_1 - (1 + \delta) p_2$, namely $q_1 = 1/2$ and $q_2 = 0$. Hence $\Pi_o = (1 + \delta)/4$, $S_o = (1 + \delta)/8$ The open loop equilibrium consists of concentrating the production and the sales on the first period, through a pricing rule which makes the residual demand of the consumers equal to zero in the second period. The result would be the same if the consumers were merged in an unique entity in charge of maximizing the aggregate surplus. In terms of profit and surplus, the open loop equilibrium is equivalent to the leasing solution (cf. Tirole, 1988, p.81)

2.2. The closed loop equilibrium

The strategies held by the players are requested to be equilibrium strategies of any subgame starting in period 2, as the consumers and the producer are allowed to observe, at the beginning of this period, the quantities sold in period 1.

At period 2, the strategy of consumer i results from the following program

$$\max_{q_2^i} \left(u \left(q_1^i + q_2^i \right) - p_2 q_2^i \right), \tag{2.12}$$

hence relation (2.6) which defines the feedback rule of consumer i $q_2^i = q_2^i \left(q_1^i, p_2 \right)$. The strategy of the producer is determined by the program :

$$
\begin{aligned}
&\max_{p_2, q_2^i} p_2 q_2 \\
&s.t \\
&u' \left(q_1^i + q_2^i \right) - p_2 = 0 \\
&i = 1, ..., n.
\end{aligned}
\tag{2.13}
$$

Necessary conditions of program (2.13) give :

$$p_2 + \ell_2^i u'' \left(q_1^i + q_2^i \right) = 0, \forall i = 1, ..., n. \tag{2.14}$$

---

[2] i.e. the bilateral monopoly between the producer and one consumer

$$\sum_{i=1}^{n} \ell_2^i = \sum_{i=1}^{n} q_2^i. \tag{2.15}$$

Relations (2.6) and (2.14) imply that $(q_1^i + q_2^i) = u'^{-1}(p_2)$ and multiplier ℓ_2^i do not depend on index i ; accordingly relation (2.15) gives $\ell_2^i = q_2/n$, and $q_2 = nu'^{-1}(p_2) - q_1$. Hence the feedback rule of the producer is given by a function of the aggregate sales in period 1, $p_2 = p_2(q_1)$, as the solution of the equation :

$$p_2 + \left(u'^{-1}(p_2) - q_1/n\right) u'' \left(u'^{-1}(p_2)\right) = 0. \tag{2.16}$$

In the linear quadratic case, the second period strategies of the consumers and the producer are given by :

$$\begin{aligned} q_2^i &= q_2^i(q_1^i, p_2) = (1 - p_2)/n - q_1^i \\ p_2 &= p_2(q_1) = (1 - q_1)/2 \end{aligned} \tag{2.17}$$

At period 1, consumer i 's strategy results from the following program :

$$\max_{q_1^i} \left[\left(u\left(q_1^i\right) - p_1 q_1^i\right) + \delta \left(u\left(q_1^i + q_2^i\left(q_1^i, p_2(q_1)\right)\right) - p_2(q_1) q_2^i\right) \right] \tag{2.18}$$

Necessary conditions give :

$$u'\left(q_1^i\right) - p_1 + \delta u'\left(q_1^i + q_2^i\right) - \delta q_2^i \partial p_2/\partial q_1 = 0. \tag{2.19}$$

or :

$$p_1 = u'\left(q_1^i\right) + \delta p_2 - \delta q_2^i \partial p_2/\partial q_1 = 0. \tag{2.20}$$

Relation 2.20)differs from relation (2.7) by the term $\delta q_2^i \partial p_2/\partial q_1$ which measures *the oligopsony effect* : if consumer i decides in the first period to postpone his purchase, this will increase the price prevailing in the second period, because of the adjustment of the producer to this demand shift. This effect was ignored in the open loop model. In the linear-quadratic case, relation (2.20) becomes $p_1 = 1 - nq_1^i + \delta p_2 + \delta q_2^i/2 = 0$. or :

$$q_1^i = \frac{4n(1 - p_2) + \delta\left(2n + 1 - \sum_{j \neq i} q_1^i(2n - 1)\right)}{4n^2 + \delta(4n - 1)}, \tag{2.21}$$

Relation (2.21) defines the best-response function of consumer i once price p_1 is quoted. Crossing these functions leads to the symmetric solution where :

$$nq_1^i = q_1(p_1) = \frac{4n(1-p_1) + \delta(2n+1)}{(4n + \delta(2n+1))}, \forall i = 1, ..., n$$

Accordingly, the producer strategy at period 1 results from the following program :

$$\max_{p_1} [p_1 q_1(p_1) + \delta p_2(p_1) q_2(p_2(q_1(p_1)), q_1(p_1))] \tag{2.22}$$

In the linear-quadratic case, we get $p_1^* = [\delta(2n+1) + 4n]^2 / 8n(\delta(n+1) + 4n)$ and then $q_1^* = (\delta + 4n)/2[\delta(n+1) + 4n]$, $p_2^* = [\delta(2n+1) + 4n]/4[\delta(n+1) + 4n] = q_2^*$. As a result :

$$\Pi_c = [\delta(2n+1) + 4n]^2 / 16n(\delta(n+1) + 4n), \tag{2.23}$$

$$S_c = \frac{\delta^3(2n+1)(2n^2 + n - 2) + 4\delta^2 n(4n^2 - 2n - 3) + 16\delta n^3 + 64n^3}{32n[\delta(n+1) + 4n]^2}. \tag{2.24}$$

Straightforward computations prove that :

$$\Pi_c \geq \Pi_o, \forall n < \left(\sqrt{1+\delta} + 1\right)/2 \tag{2.25}$$

$$S_c \leq S_o, \forall n \in N. \tag{2.26}$$

In addition, the selling solution by Bulow and Tirole is found as the limiting case when $n \to +\infty$. For instance, $\lim_{n \to \infty} \Pi_s = (2+\delta)^2 / 4(4+\delta)$. Consequently, the producer is better off when selling the durable good for low values of the number of customers. The more oligopsonic is the market, the more profitable is the selling of the durable. Of course, this result is obtained in the linear-quadratic case ; it has to be validated for more general demand structures.

3. A differential game approach of the continuous time problem

Extending the two-period previous model results in a *leader-follower* differential game between the producer and the consumers. Such an approach has been developed by many authors to cope with multi-level continuous time games (Basar & Haurie, 1982, Wishart & Olsder, 1979, Thépot, 1990)

3.1. Statement of the model

We consider now that the producer faces the set of n consumers over an infinite horizon $[0, +\infty[$. In period $[t, t + dt[$, the producer charges a price, $p(t)$, while consumer i purchases a quantity $q_i(t)\, dt$. Let $Q_i(t) = Q_{0i} + \int_{s=0}^{t} q_i(s)\, ds$, the cumulated sales of consumer i, where Q_{0i} stands for the stock of good owned at time 0 by consumer i through previous purchases. Let $q(t) = \sum_{i=1}^{n} q_i(t)$ and $Q(t) = \sum_{i=1}^{n} Q_i(t)$, the aggregate sales (resp. cumulated sales) at time t. Let $S > Q_0$ the maximum size of the market. We shall assume that the individual utility function u per unit of time is the same for all the consumers; function u is C^2-differentiable, strictly increasing and concave on $[0, S/n[$ and constant elsewhere. Let $f : Q \in \Re^+ \to u'(Q/n) \in \Re^+$ be the *inverse aggregate rental demand*, with $f > 0$ on $[0, S[$, $f = 0$ on $[S, +\infty[$. Function f is still assumed to be independent on n. As previously, there is no production cost incurred by the producer. Let r be the discount rate, which is assumed to be identical for the producer and the consumers. The problem can then be stated as the following differential game :

$$
\begin{cases}
leader \to \; \max_{p} \int_{0}^{+\infty} p(t)\, q(t)\, e^{-rt} dt, \\[2mm]
followers \to \max_{q_i} \int_{0}^{+\infty} [u(Q_i(t)) - p(t)\, q_i(t)]\, e^{-rt} dt,
\end{cases}
$$

$$
\frac{dQ_i}{dt}(t) = q_i(t), Q_i(0) = Q_{0i}, \tag{3.1}
$$

$$
p(t) \geq 0. \tag{3.2}
$$

As in the two-period case, three types of equilibria may be considered :

- *The open loop equilibrium*, where the strategies are defined by $p = p(t)$, $q_i = \xi_i(p(.))$.

- *The closed loop equilibrium*, where the strategies are defined by $p = \pi(Q, t)$, $q_i = \zeta_i(\pi(.,.))$.

For simplicity, we will characterize these equilibria without price condition (3.2), which will be treated a posteriorly.

3.2. The open-loop equilibrium

The consumers have to determine their purchase programs over $[0, +\infty[$ as the optimal response to any pricing program $\{p(t)\}$ where function p is assumed to be C^2-differentiable.

3.2.1. Necessary conditions of optimality of consumers' strategy

Consumer i faces a standard optimal control problem, where q_i is the control variable and Q_i the state variable :

$$
\begin{cases}
\max \int_0^{+\infty} \left[u\left(Q_i\left(t\right)\right) - p\left(t\right) q_i\left(t\right) \right] e^{-rt} dt \\
\dot{Q}_i\left(t\right) = q_i\left(t\right), Q_i\left(0\right) = Q_{0i}; \\
\qquad q_i\left(t\right) \geq 0 \\
\qquad p\left(.\right), fixed.
\end{cases}
\tag{3.3}
$$

Applying the Maximum Principle results in the following necessary conditions, where λ_i stands for the costate variable associated with variable Q_i and γ_i the multiplier associated with the positivity constraint $q_i\left(t\right) \geq 0$:

$$
\dot{\lambda}_i = r\lambda_i - u'\left(Q_i\right)
\tag{3.4}
$$

$$
\lambda_i - p + \gamma_i = 0, \gamma_i \geq 0, \gamma_i q_i = 0. \ ,
\tag{3.5}
$$

with the transversality condition :

$$
\lim_{t \to +\infty} \lambda_i\left(t\right) e^{-rt} = 0.
\tag{3.6}
$$

The optimal strategy of consumer i is based on the combination of two regimes according to the value of multiplier γ_i :

- *The no purchase regime*, where $q_i = 0$, and $p > \lambda_i$.

- *The purchase regime*, where $q_i \geq 0$, $p = \lambda_i$. then $\dot{p} = rp - u'\left(Q_i\right)$. Hence :

$$
q_i\left(t\right) = - \left[\ddot{p}\left(t\right) - r\dot{p}\left(t\right)\right] / \left[u''\left(u'^{-1}\left(rp\left(t\right) - \dot{p}\left(t\right)\right)\right)\right]
\tag{3.7}
$$

Accordingly the purchase regime is feasible if :

$$
\ddot{p}\left(t\right) - r\dot{p}\left(t\right) \geq 0, \dot{p}\left(t\right) - rp\left(t\right) \leq 0
\tag{3.8}
$$

We will a priori assume that the pricing function $p(t)$ likely to be used by the producer satisfies condition (3.8).

The program of the producer amounts to determine function $p^*(.)$ so as to maximize the profit, $\int_0^{+\infty} p(t)q(t)e^{-rt}dt$ knowing that $\lambda_i = p$, namely :

$$\dot{p} = rp - f(Q). \tag{3.9}$$

Hence the following optimal control problem :

$$\begin{cases} \max \int_0^{+\infty} p(t)q(t)e^{-rt}dt, \\ \dot{p} = rp - f(Q), \lim_{t \to +\infty} p(t)e^{-rt} = 0; \\ \dot{Q}(t) = q(t), Q(0) = Q_0; \end{cases} \tag{3.10}$$

where q is the control variables, p and Q the state variables. Necessary conditions are the following, where α and β stand for the costate variables associated respectively with p and Q :

$$\dot{\alpha} = -q, \tag{3.11}$$

$$\dot{\beta} = r\beta + \alpha f'(Q), \tag{3.12}$$

$$p + \beta = 0. \tag{3.13}$$

Note that variable p faces a terminal condition (3.6). The transversality condition related to costate variable α resorts to the initial condition $\alpha(0) = 0$, then , by relation (3.11), $\alpha = Q_0 - Q$. Accordingly, relations (3.12) and (3.13) yield :

$$\dot{p} = rp + (Q - Q_0)f'(Q), \tag{3.14}$$

Relations (3.9) and (3.14) are satisfied along the equilibrium path if :

$$f(Q) + (Q - Q_0)f'(Q) = 0. \tag{3.15}$$

Let $Q^m(Q_0)$ be the (unique) solution of equation (3.15) ; quantity $\Delta^m = Q^m(Q_0) - Q_0$ is the aggregate quantity purchased in a single-period monopoly where Q_0 is already held by the consumers at time 0, i.e. $\Delta^m = \arg\max_{\Delta} f(Q_0 + \Delta)\Delta$. The equilibrium strategy of the producer is given by :

$$p^*(t) = p^m(Q_0), \tag{3.16}$$

where $p^m(Q_0) \equiv f[Q^m(Q_0)]/r$. Hence the equilibrium path is determined as follows : at time 0, the consumers instantaneously purchase quantity Δ^m and does not purchase anything further, i.e. $q_i^*(t) = 0, \forall t > 0$; the producer charges price $p^m(Q_0)$ for ever. Clearly, the open loop equilibrium suffers from strong *time inconsistency*, since the strategy of the producer depends on the initial state, Q_0, of the game : if at time $\epsilon > 0$ for instance, the producer decides to update his strategy according to rule (3.16), the equilibrium path is changed since the price becomes $p^m(Q^m(Q_0)) < p^m(Q_0)$, with a new jump of purchase orders, and so on at time $2\epsilon, 3\epsilon, \dots$; then price $p^m \to 0, \forall \epsilon$; this is typically the key argument of the Coase conjecture : at any time, the producer is tempted to decrease his price so as to capture the residual demand.

3.3. The closed loop equilibrium

The price charged by the producer at time t depends now on the cumulated sales . In this case we are looking for time-consistent strategies $\pi(Q)$ and $q(Q)$ respectively of the producer and the consumers. Consumer i's program is similar to (3.3) where price p is considered as given by the feedback rule $\pi(Q,t)$ of the producer. Since the dynamics of the system is autonomous and the horizon infinite, the feedback rule may be a priori restricted to be a mere function of the cumulated sales, i.e. $\pi(Q)$. Furthermore by only considering symmetric solutions as previously, necessary conditions (using the same notations) yield :

$$\dot{\lambda} = r\lambda - f(Q) + (q/n)\,\pi'(Q), \tag{3.17}$$

$$\lambda - \pi + \gamma = 0, \gamma \geq 0, \gamma q = 0, \tag{3.18}$$

$$\lim_{t \to +\infty} \lambda(t)\,e^{-rt} = 0. \tag{3.19}$$

As in the open loop case two situations may occur :

- *The no purchase policy* with $q = 0, \pi > \lambda$

When it is used over the whole horizon, this policy deals with Q constant and $\lambda = f(Q)/r$.

- *The purchase policy* with $q \geq 0, \pi = \lambda$

Hence relation (3.17) become $\left(\frac{n-1}{n}\right) q(t) \pi'(Q(t)) = r\pi(Q(t)) - f(Q(t))$. As we are looking for *time-consistent strategy namely the strategies have to be defined even outside the optimal trajectory*, the purchase strategy of the consumers is defined, for any value, Q and for any pricing rule, $\pi(.)$ by the relation :

$$\left(\frac{n-1}{n}\right) q\pi'(Q) = r\pi - f(Q) \tag{3.20}$$

or, in feedback form :

$$q[Q, \pi(.)] = \frac{n(r\pi - f(Q))}{(n-1)\pi'(Q)} \tag{3.21}$$

The purchase policy is not defined for $n = 1$; it is feasible only if it leads to a trajectory such that :

$$\lim_{t \to +\infty} \pi(Q(t)) e^{-rt} = 0. \tag{3.22}$$

The time-consistent consumer strategy can be described as follows, for Q given :

- if $\pi(Q) \geq f(Q)/r$, the consumers do not purchase the durable, $q = 0$,
- if $\pi(Q) < f(Q)/r$ and $\partial\pi/\partial Q < 0$, the purchased quantity, q, is positive and given by relation (3.20).
- if $\pi(Q) < f(Q)/r$ and $\partial\pi/\partial Q \geq 0$, the consumer strategy does not exist.

- For $n = 1$, only the no purchase strategy can be used by the consumers, whatever the pricing policy used by the monopolist is.

- The pricing strategy of the producer for $n > 1$ results from the following program :

$$\begin{cases} \max_{\pi(.),q,Q} \int_0^{+\infty} \pi(Q(t)) q(t) e^{-rt} dt \\ q(t)\pi'(Q(t)) = \left(\frac{n}{n-1}\right) [r\pi(Q(t)) - f(Q(t))], \\ \dot{Q}(t) = q(t), \\ Q(0) = Q_0, \\ \lim_{t \to +\infty} \pi(Q(t)) e^{-rt} = 0 \end{cases} \tag{3.23}$$

It is given by the following proposition which expresses the Coase conjecture :
Proposition 1

- *The optimal pricing strategy of the producer is given by* :

$$\pi^* (Q) = 0, \forall Q \tag{3.24}$$

- *The optimal purchase strategy of the consumers is given by* :

$$q^* (Q) = +\infty \text{ for } Q < S, \ q^* (S) = 0. \tag{3.25}$$

- *The optimal trajectory is defined by* :

$$\begin{aligned} Q^* (t) = S, \forall t > 0 \\ q^* (0) = +\infty, q^* (t) = 0, \text{ for } t > 0 \end{aligned} \tag{3.26}$$

Proof : see appendix 1 ■.

It is worth mentioning that the solution does not depend on initial condition $Q(O) = Q_0$, as the equilibrium strategies are time-consistent.

4. Concluding remarks

Our results suggest that the oligopsony effect might have a positive impact on the monopoly profit in discrete time models when the number of customers is low ; however it turns out not to be strong enough to offset the intertemporal price discrimination in the continuous time case. But in discrete time models, one might expect to see the profit of a durable-goods monopolist decreasing with the number of consumers (for a given aggregate demand) ; in other words, the monopsony power of the buyers could restrain the intertemporal price discrimination and enhance the profit of the firm. The continuous time model indicates how differential games techniques can be fairly applied to solve durable-goods problems ; it remains to use them in more general situations, particularly when the consumers preferences are assumed to be heterogeneous.

References

[1] Bagnoli,M., S.W. Salant and J.E. Swierzbinsky, (1989), "Durable-goods monopoly with discrete demand", *Journal of Political Economy*, **97**, 1459-1478.

[2] Basar, T. and A. Haurie, (1982), "Feedback-Stackelberg solution in continuous time muli-level games" in *Proceedings of the 21th conference on Decisions & Control* (Orlando FL), 664-668.

[3] Basar, T. and G. Olsder, (1982), *Dynamic noncooperative game theory*, Academic Press, London.

[4] Bulow, J., (1982), "Durable-goods monopolists", *Journal of Political Economy*, 90, 314-332

[5] Coase, R., (1972),"Durability and monopoly", *Journal of Law and Economics*, 15,143-149

[6] Gul, F., Sonnenschein, H. and Wilson, R., (1986), "Foundations of dynamic monopoly and the Coase conjecture", *Journal of Economic Theory*, 39, 155-190

[7] Dudey, M.(1993), "On the foundations of dynamic monopoly theory", mimeo Rice University forthcoming in the *Journal of Political Economy*

[8] Sobel, J., (1991), "Durable-goods monopoly with entry of new consumers", *Econometrica*, 59, 1455-1485

[9] Stokey, N.L., (1981), "Rational expectations and durable goods pricing", *Bell Journal of Economics*, 12, 112-128

[10] Thépot, J., (1990), "Nash versus Stackelberg strategies in a capital accumulation game", in *Analysis and Optimization of Systems*, A. Bensoussan and J.L. Lions eds., Springer Verlag, Berlin, 735-744

[11] Tirole, J., (1988), *The Theory of Industrial Organization*, MIT-Press

[12] Van Cayseele, P., (1991), "Consumer rationing and the possibility of intertemporal discrimination", *European Economic Review*, 1473-1484

[13] Wishart D. and G. Olsder, (1979), "Discontinuous Stackelberg solution", *International Journal of Systems Science*, 10, 1359-1368.

Appendix 1: Proof of Proposition 1

The proof is given using infinite-dimensional optimization techniques. Let us define the Lagrangian of problem (3.23) as, where Q, β and q are functions of t :

$$\Delta = \int_0^{+\infty} \left\{ \begin{array}{c} \pi(Q)q + \\ \alpha(Q)\left[-q\pi'(Q) + (r\pi(Q) - f(Q))\left(\frac{n}{n-1}\right)\right] + \beta\left[q - \dot{Q}\right] \end{array} \right\} e^{-rt}dt \quad (.1)$$
$$-\beta(0)\left[Q(0) - Q_0\right]. \quad (.2)$$

Integrating by parts, we have :

$$\int_0^{+\infty} \alpha(Q)\left[q\pi'(Q)\right]e^{-rt}dt = \left[\alpha(Q)\pi(Q)e^{-rt}\right]_0^{+\infty} \quad (.3)$$

$$-\int_0^{+\infty} \pi(Q)\left[-r\alpha(Q) + \alpha'(Q)q\right]e^{-rt}dt. \quad (.4)$$

and :

$$\int_0^{+\infty} \beta\dot{Q}e^{-rt}dt = \left[\beta Qe^{-rt}\right]_0^{+\infty} - \int_0^{+\infty} Q\left(\dot{\beta} - r\beta\right)e^{-rt}dt. \quad (.5)$$

Hence :

$$\Delta = \int_0^{+\infty} \left\{ \begin{array}{c} \pi(Q)q + \pi(Q)\left[-r\alpha(Q) + \alpha'(Q)q\right] + \\ \alpha(Q)(r\pi(Q) - f(Q))\left(\frac{n}{n-1}\right) \end{array} \right\} e^{-rt}dt +$$
$$+ \int_0^{+\infty} \left[Q\left(\dot{\beta} - r\beta\right) + \beta q\right]e^{-rt}dt - \beta(0)\left[Q(0) - Q_0\right]. + \alpha(Q(0))\pi(Q(0)), \quad (.6)$$

because of relation (3.22) and if the following transversality condition is satisfied

$$\lim_{t \to +\infty} \beta(t)e^{-rt} = 0.$$

Maximizing Δ for $t = 0$, i.e. with respect to $\pi(Q(0))$ and $Q(0)$ yields :

$$\alpha(Q(0)) = 0, \alpha'(Q(0)) \text{ or } \pi(Q(0)) = 0 \quad (.7)$$

For $t > 0$, necessary conditions are found by differentiating the integrand of (.6) with respect to π, q and Q , namely :

$$q + \alpha'(Q) q + \frac{r\alpha(Q)}{n-1} = 0; \tag{.8}$$

$$\pi(Q)[1 + \alpha'(Q)] + \beta = 0. \tag{.9}$$

$$\pi'(Q)\left[q + \alpha'(Q)q + \frac{r\alpha(Q)}{n-1}\right] + \pi(Q)\left[\frac{r\alpha'(Q)}{n-1} + \alpha''(Q)q\right] - \\ - \left(\frac{n}{n-1}\right)\frac{d}{dQ}[\alpha(Q)f(Q)] + \left(\dot{\beta} - r\beta\right) = 0. \tag{.10}$$

Relation (.9) gives :

$$\dot{\beta} - r\beta = -\left[\pi(Q)\alpha''(Q) + \pi'(Q)(1 + \alpha'(Q))\right]q + r\pi(Q)[1 + \alpha'(Q)] \tag{.11}$$

Hence relation (.10) gives :

$$\pi'(Q)r\alpha(Q)/(n-1) + r\pi(Q)\left[1 + \left(\frac{n}{n-1}\right)\alpha'(Q)\right] - \\ - \left(\frac{n}{n-1}\right)\frac{d}{dQ}[\alpha(Q)f(Q)] = 0 \tag{.12}$$

We are looking for a solution leading to functions α, π which remain C^1-differentiabl along the optimal path, actually continuous at time 0 with initial conditions (.7). Clearly conditions (.8) and (.12 are satisfied for :

$$q(O) = +\infty, q(t) = 0; Q(t) = S, \forall t > 0, \text{ and } \pi(Q(t)) = \alpha(Q(t)) = 0, \forall t \geq 0. \tag{.13}$$

The **time-consistent** strategies of the producer and the consumers are then given by relations (3.25) and (3.26). Hence proposition 1[3].

---

[3]Other strategies which do not meet the time-consistency conditions are not excluded

Government Price Subsidies to Promote Fast Diffusion of a New Consumer Durable

Engelbert J. Dockner[1], Andrea Gaunersdorfer[1] and Steffen Jørgensen[2]

[1] University of Vienna, Austria
[2] Odense University, Denmark

Abstract. We consider a market in which a single supplier sells a new product characterized by diffusion effects on the demand side. In this setting we analyze the problem whether or not the government should subsidize the diffusion of this innovation. We assume that the government is a Stackelberg leader and decides about the subsidy (price or cost subsidy) before the supplier sets his optimal price. The objective of the government is to choose a subsidy policy such that the number of adopters at the horizon date is maximized. The firm chooses a pricing strategy so as to maximize the present value of profits.

1 Introduction

Promoting fast diffusion of a new product is not necessarily the first-best policy as social welfare is concerned. If such an enterprise is too costly, a policy promoting fast diffusion could turn out to be welfare-decreasing. To settle this problem, a basic difficulty lies in specifying a welfare function that takes into account the effects on welfare of both *fast diffusion* and the *cost* of achieving this objective. Only if such a function can be established it is possible to assess properly the trade off between fast diffusion and cost.

Without a welfare function, one approach is to evaluate the impacts of various policy instruments on the innovating firm's objective, tacitly assuming that accelerated diffusion is socially beneficial. Kalish and Lilien (1983) and Zaccour (1995) studied government price subsidies of a new consumer good, considering both a durable and a repeat purchase good. The latter case is perhaps less interesting since in real life most examples of government subsidies are found in durable goods. In Kalish and Lilien (1983) the problem was cast as an optimal control problem of a government who wishes to choose a time path for its price subsidy to the consumers. The good was produced and sold by a single firm (assuming, for instance, the firm has a monopoly based on patent protection). The objective of the government was to maximize the cumulative number of units sold by some fixed instant of time, given a predetermined budget. Zaccour (1995) studied this problem as a Nash game between the government and the firm, assuming that both players precommit to open-loop strategies throughout the game.

In this paper we are concerned with a monopolistic firm producing a durable good. We formulate the problem as a Stackelberg game with the government as the leader. This approach was suggested (but not analyzed) in Kalish and Lilien (1983) and Zaccour (1995). In order to capture strategic interactions we assume that both players use feedback strategies. We discuss different subsidy policies and analyze the game for various price response functions. Not surprisingly, it turns out that the optimal policy for a government which has no budget constraint is to give maximal subsidy if the number of units sold by the end of the planning horizon is to be maximized.

We describe the different models in Section 2. In Section 3 we present our results for feedback strategies and in Section 4 we discuss open-loop strategies. Section 5 concludes.

2 The Models

Consider the case of a new durable, produced and sold by a single firm, and suppose that the government, for some reason, wishes to accelerate the penetration of this product. Assume each consumer only buys at most one unit of the product during the planning period. Let t denote time, $0 \leq t \leq T$, where T is the time span for the government's subsidizing activities. Represent by $x(t)$ the number of adopters by time t (the cumulative sales by time t). Thus $\dot{x}(t)$ represents the sales rate at time t. By $p(t)$ we denote the price per unit obtained by the firm at time t. Assume production is equal to sales. For simplification we take the unit cost of the firm c as constant. (Kalish and Lilien (1983), Zaccour (1995) introduce cost experience effects.)

For the subsidy scheme, consider the following alternatives.

Model I: The government pays *the firm* a subsidy $\sigma = \sigma(t)$ (a dollar amount). The subsidy decreases the firm's unit cost (or acts as a markup on price). Thus the firm's margin becomes $p - (c - \sigma) = (p + \sigma) - c$. The firm quotes price p on which consumer demand reacts.

Model II: The firm quotes price p but the consumer only pays $p - \beta$. The *firm* gets a subsidy of $\beta = \beta(t)$ dollars from the government. An alternative scenario is as follows: the consumer pays price p to the firm and the government pays β to the *consumer*, i.e. the latter still pays a net price of $p - \beta$. Consumer demand reacts on net consumer price $p - \beta$.

The main difference between the two models lies in the fact that consumer demand reacts only on p (the firm's control) in Model I, whereas consumer demand reacts upon p and β (firm's and government's control) in Model II.

Model III: Let $\gamma = \gamma(t)$ denote the fraction of p subsidized by the government. The buyer pays $(1 - \gamma)p$ to the firm and the government pays a subsidy γp to the *firm*. Thus, the firm gets price p. Alternatively, a buyer pays price p to the firm and the *buyer* receives a refund γp from the government, i.e.

the buyer pays a net price of $(1-\gamma)p$. Consumer demand reacts on $(1-\gamma)p$ and therefore on both control variables.

Note the similarity between Models II and III. In both models the firm cashes price p, the consumer pays $p-\beta$ resp. $p-\gamma p$, and the government's expense is β resp. γp. Consumer demand reacts on $p-\beta$ resp. $p-\gamma p$, i.e. on both controls. A technical difference between the two models lies in the fact that the controls enter additively in Model II, multiplicatively in Model III.

State equation for x

We employ a simple new product diffusion model. The sales rate of the durable good, $\dot{x}$, depends on cumulative sales and the appropriate price paid by the consumer to the firm, P say $(P \in \{p, p-\beta, (1-\gamma)p\})$, and is specified as

$$\dot{x}(t) = f(P(t))(N - x(t)). \tag{1}$$

$N = \text{const.} > 0$ is a fixed market potential and $f'(P) < 0$. The model is a Bass model without an imitation term and a price-dependent coefficient of innovation. Thus the assumption is that diffusion is driven by innovators only. Note, x and P are multiplicatively separable and $\partial\dot{x}/\partial x < 0$ for all x, that is, we have a model with saturation effect only.

For the price response function f various specifications are available. For our analysis we consider the following three cases.

Case 1: Linear price response function (Eliashberg and Jeuland (1986), Dockner and Gaunersdorfer (1995))

$$f(P) = \alpha - kP. \tag{2}$$

Case 2: Exponential price response function (Kalish and Lilien (1983), Zaccour (1995))

$$f(P) = e^{-kP}. \tag{3}$$

Case 3: Price response function with constant elasticity $-\eta$,

$$f(P) = (kP)^{-\eta}, \quad \eta > 1. \tag{4}$$

Objective functions

We assume that government maximizes penetration by time T, i.e., $x(T)$, subject to state equation (1) (cf. Kalish and Lilien (1983), Zaccour (1995)). The time horizon T is fixed and finite. Thus, the objective function of the government is given as

$$J_G = x(T) = \int_0^T \dot{x}(t)\,dt + x_0. \tag{5}$$

The firm maximizes the present value of its profit stream over the time interval $[0, T]$,

$$J_F = \int_0^T e^{-rt}(\pi(t) - c)\dot{x}(t)\,dt. \tag{6}$$

Here, $\pi - c$ denotes the appropriate unit margin,

$$\pi(t) - c = \begin{cases} p(t) + \sigma(t) - c & \text{in Model I} \\ p(t) - c & \text{in Models II and III} \end{cases}$$

and r is the firm's discount rate.

Our differential game becomes

$$\max_{s \in [0,\bar{s}]} J_G = \max_{s \in [0,\bar{s}]} \int_0^T \dot{x}(t)\, dt$$

$$\max_{p \geq c} J_F = \max_{p \geq c} \int_0^T e^{-rt}(\pi(t) - c)\dot{x}(t)\, dt$$

subject to

$$\dot{x}(t) = f(P(t))(N - x(t))$$
$$x(0) = x_0 = 0.$$

In J_G, $s \in \{\sigma, \beta, \gamma\}$ denotes the subsidy. Let us assume that σ resp. β are bounded from above by a fixed constant, i.e. $\sigma \in [0, \bar{\sigma}]$ ($\bar{\sigma} \leq c$), $\beta \in [0, \bar{\beta}]$, thus $\bar{s} \in \{\bar{\sigma}, \bar{\beta}, 1\}$. The government declares its strategy first and the firm reacts to it. Hence we are dealing with a Stackelberg game with the government as leader. We are interested in both feedback and open-loop Stackelberg equilibria (see, e.g. Başar and Olsder (1995)).

3 Feedback Stackelberg Equilibrium Strategies

We first compute the instantaneous best response of the firm to a given subsidy strategy of the government and then calculate the optimal subsidy strategy of the government. The latter knows the firm's best response function. We derive the equilibrium strategies by using a dynamic programming approach. Let

$$V_F(x, t) = \max_{p \geq c} \int_t^T e^{-r\tau}(\pi - c)\dot{x}\, d\tau$$

$$= \max_{p \geq c} \int_t^T e^{-r\tau}(\pi - c)f(P)(N - x)\, d\tau \qquad (7)$$

$$V_G(x, t) = \max_{s \in [0,\bar{s}]} \int_t^T \dot{x}\, d\tau = \max_{s \in [0,\bar{s}]} \int_t^T f(P)(N - x)\, d\tau \qquad (8)$$

denote the optimal value functions for the firm and the government.

Suppose that the value functions are continuously differentiable in both arguments. Then the value functions have to satisfy the Hamilton-Jacobi-Bellman equations

$$-\frac{\partial V_F}{\partial t} = \max_{p \geq c}\left\{ \left(e^{-rt}(\pi - c) + \frac{\partial V_F}{\partial x} \right) f(P)(N - x) \right\} \qquad (9)$$

$$-\frac{\partial V_G}{\partial t} \;=\; \max_{s \in [0,\bar{s}]} \left\{ \left(1 + \frac{\partial V_G}{\partial x}\right) f(P(s))(N - x) \right\} \tag{10}$$

as well as the transversality conditions

$$V_F(x, T) \;=\; 0 \tag{11}$$
$$V_G(x, T) \;=\; 0. \tag{12}$$

Note, that P depends upon s since government declares its policy s first and the firm sets its strategy p accordingly.

Theorem 1 *For all three price response functions, the optimal strategy of the government is to give maximal subsidy of $s^*(t) \equiv \bar{s}$, i.e. $\sigma^*(t) \equiv \bar{\sigma}$ (Model I), $\beta^*(t) \equiv \bar{\beta}$ (Model II), $\gamma^*(t) \equiv 1$ (Model III).*

PROOF: In what follows we use the $*$-notation for optimal strategies. To obtain the optimal strategies we first determine the best response p^* to any given s and then derive the optimal strategy s^*.

Maximizing the right hand side of (9) with respect to p gives, for $p^* > c$,

$$f(P^*)\left.\frac{d\pi}{dp}\right|_{p=p^*} + f'(P^*)\left.\frac{dP}{dp}\right|_{p=p^*}(\pi^* - c) + e^{rt}f'(P^*)\left.\frac{dP}{dp}\right|_{p=p^*}\frac{\partial V_F}{\partial x} = 0. \tag{13}$$

Note $d\pi/dp = 1$ and $dP/dp \in \{1, 1, -\gamma\}$ for price response functions (2)–(4).

Assume that the value functions are as follows:

$$V_F(t, x(t)) \;=\; a(t)(N - x(t)) \tag{14}$$
$$V_G(t, x(t)) \;=\; b(t)(N - x(t)), \tag{15}$$

where $a(t)$ and $b(t)$ are continuously differentiable, non-negative real-valued functions. Then the following pricing strategies are optimal for price response functions (2), (3), and (4):

Model I

$$(2): \quad p^*(\sigma) \;=\; \frac{1}{k+1}(\alpha + c - \sigma + e^{rt}a) \tag{16}$$

$$(3): \quad p^*(\sigma) \;=\; \frac{1}{k} + c - \sigma + e^{rt}a \tag{17}$$

$$(4): \quad p^*(\sigma) \;=\; \frac{\eta}{\eta - 1}(c - \sigma + e^{rt}a) \tag{18}$$

Model II

$$(2): \quad p^*(\beta) \;=\; \frac{1}{k+1}(\alpha + c + \beta + e^{rt}a) \tag{19}$$

$$(3): \quad p^*(\beta) \;=\; \frac{1}{k} + c + e^{rt}a \tag{20}$$

$$(4): \quad p^*(\beta) \;=\; \frac{\eta}{\eta - 1}(c + \frac{1}{\eta}\beta + e^{rt}a) \tag{21}$$

Model III

$$(2): \quad p^*(\gamma) = \frac{1}{2}\left(\frac{\alpha}{k(1-\gamma)} + c + e^{rt}a\right) \tag{22}$$

$$(3): \quad p^*(\gamma) = \frac{1}{k(1-\gamma)} + c + e^{rt}a \tag{23}$$

$$(4): \quad p^*(\gamma) = \frac{\eta}{\eta-1}(c + e^{rt}a). \tag{24}$$

Note that, in Model II, the optimal pricing strategy does not depend on the subsidy policy for the exponential price response function (3), in Model III for function (4). Hence, these cases show that under certain circumstances the government cannot influence the firm's pricing strategy at all. In Model II the constraint $\beta \leq p^*$ is always fulfilled for the other cases.

To obtain the optimal subsidy policy s^*, we maximize the right hand side of (10). This is accomplished by maximizing $f(P^*(s))\mathrm{sgn}\,(1 + \frac{\partial V}{\partial x})$. From equations (10) and (15) we have

$$-\dot{b}(t) = f(P^*(s^*))(1 - b(t)).$$

Dividing by $1 - b(t)$ and integrating[1] gives

$$1 - b(t) = \kappa \exp\left(\int f(P^*(s^*))\,d\tau\right).$$

Since $b(T) = 0$, provided $N - x(T) > 0$, κ is strictly positive. Hence $1 - b(t) > 0$, which shows that $1 + \frac{\partial V_G}{\partial x} > 0$ in (10).

Thus $s^*(t) = \bar{s}$ since f is strictly decreasing in P and P^* is strictly decreasing in s (i.e., f is strictly increasing in s). (In Model III this holds only if $\eta > 2$.)

In (22) and (23) p^* is not defined for $\gamma = 1$. Since p^* has to be chosen so as to maximize the right hand side of (9),

$$-\frac{\partial V_F}{\partial t} = \max_p\left\{\left(e^{-rt}(p - c) - a\right) f(\tilde{k})(N - x)\right\}$$

($\tilde{k}$ being some constant), which is a linear function in p, it is optimal to set p at its upper bound. $\qquad\Box$

Theorem 2 *For Models I and II, and Model III with a constant elasticity price response function, and a sufficiently small discount rate r, the optimal pricing strategies are nonincreasing over time.*

PROOF: According to (16)–(21) and (24), the time derivative of the equilibrium price is given as

$$\dot{p}^*(t) = \nu e^{rt}(ra(t) + \dot{a}(t)),$$

---

[1] We assume that $1 - b(t) \neq 0$, i.e. $V_G \neq 0$. If $\exists t: 1 - b(t) = 0$, the right hand side of (10) is maximized by *any* s.

where ν is some positive constant. From (9), (16)–(21), and (24) it follows that $-\dot{a}(t) < 0$. Therefore, $\dot{p}^*(t) < 0$ for r sufficiently small since $a(t)$ is assumed non-negative. $\qquad\qquad\qquad\qquad\qquad\qquad\qquad\qquad\qquad\qquad\qquad\quad\square$

The result in Theorem 2 corresponds to the one obtained by Zaccour (1995). It is a consequence of our specification of the dynamic demand function for which saturation effects dominate. Hence it resembles the intertemporal price discrimination strategies found in, for example, Kalish (1983) and Dockner and Jørgensen (1988).

4 Open-Loop Stackelberg Equilibrium Strategies

In the preceding section we derived equilibrium price and subsidy policies for the firm and the government under the assumption that the players choose feedback strategies. This implies that each player observes the current state variable (the cumulative sales volume) and designs his strategy accordingly. In this setting we showed that it is optimal for the government to choose the maximum subsidy to enhance adoption. Since the government puts emphasis on maximizing the cumulated sales by time T, it is optimal for it to provide subsidies to the consumers at the maximum amount.[2]

In this section we analyse the model under the assumption that the firm and the government choose open-loop rather than feedback strategies. This implies that both players can and will make binding commitments over the entire planning horizon and that they design at time zero their strategies as simple time paths. The assumption of irrevocable commitment removes the government's incentive to reoptimize at later instants of time. We identify an open-loop equilibrium in order to analyze how sensitive our results are with respect to changes in the players' strategy spaces. We restrict our analysis to Model III: The analysis of Models I and II can be done along the same lines as for Model III.

We assume that the firm faces a linear price response function given by

$$f(p) = \alpha - k(1 - \gamma)p.$$

In order to guarantee a non-negative profit rate as well as a non-negative sales rate we restrict the producer price by $c \le p \le \frac{\alpha}{k(1-\gamma)}$. Under this assumption the state constraint $x(t) \ge 0$ is satisfied. Suppressing time arguments, the current value Hamiltonian of the firm is given by

$$H_F = (N - x)[(p - c + \lambda_F)(\alpha - k(1 - \gamma)p)],$$

---

[2] This is partly a consequence of the assumption that government does not face any budget constraint. In Section 5 we comment briefly on the effects of a budget constraint.

with boundary conditions $\lambda_G(T) = 1$ and $\mu(0) = 0$. Integrating the adjoint equations yields

$$\lambda_G(t) = \exp\left(-\int_t^T (\alpha - k(1-\gamma)p^*)\, d\tau\right) > 0$$

$$\mu(t) = -e^{-A(t)} \int_0^t \frac{1}{2}(N-x)\lambda_G k(1-\gamma)e^{A(\tau)}\, d\tau,$$

where $A(t) \equiv (r + \alpha - ck(1-\gamma))t + k(1-\gamma)\int_0^t \lambda_F(\tau)\, d\tau$.

Hence, it also follows that $\mu(t) \leq 0$. Maximization of the Hamiltonian with respect to γ, and assuming an interior solution, results in

$$\gamma_{1,2} = 1 \pm \sqrt{\Omega}, \tag{25}$$

where Ω is defined as

$$\Omega \equiv \frac{-\alpha^2\mu}{k^2[2\lambda_G(N-x) - \mu(c - \lambda_F)](c - \lambda_F)} \geq 0.$$

It follows that $\gamma_1 > 1$ whenever $\mu < 0$ and hence γ_1 is infeasible. Therefore, we analyze the solution $\gamma = 1$. From the expression for the adjoint variable μ we conclude that $\mu(t) = 0\ \forall t$. With $\mu = 0$ the Hamiltonian of the government becomes

$$H_G\mid_{\mu=0} = \frac{1}{2}\lambda_G(N-x)(\alpha - k(1-\gamma)(c - \lambda_F)),$$

which is linear and increasing in γ. Hence $\gamma = 1$ maximizes the Hamiltonian. It also holds that $\lambda_G = e^{-\alpha(T-t)} > 0$. Hence for $\gamma = 1$ we get the rather extreme result that the buyer pays a price p to the firm which is entirely refunded by the government.

Let us look at the pricing policy that goes along with such a subsidy scheme. Since it follows that $\lambda_F = 0$, the Hamiltonian of the firm

$$H_F = (N-x)(p - c + \lambda_F)\alpha$$

is maximized whenever p is set at its maximum level possible.

What remains to be shown is that a subsidy policy with $\gamma < 1$ is not optimal. Note that if we have $\gamma < 1$ it holds that

$$\frac{\partial \dot{x}}{\partial \gamma} = \frac{1}{2}k(N-x)(c - \lambda_F) > 0.$$

Suppose we use $\gamma_2 < 1$ from (25) on $[T-\epsilon, T]$. Choosing another control $\gamma_2 + \delta(t)$, $\delta(t) > 0$, increases $\dot{x}$ and hence makes $x(T)$ larger than the one obtained for γ_2. Hence, this argument leads to the choice $\gamma_2 + \delta(t) = 1$, i.e. $\gamma(t) = 1$. Since the argument can be repeated for all t, it leads to $\gamma^* \equiv 1$ as the optimal subsidy policy.

The results for the open-loop Stackelberg equilibrium are summarized in the following Theorem.

where λ_F is the current value adjoint variable of the firm. An interior pricing policy satisfies

$$\frac{\partial H_F}{\partial p} = (N - x)[-(p - c + \lambda_F)k(1 - \gamma) + \alpha - k(1 - \gamma)p] = 0.$$

Hence for the case $\gamma < 1$ the optimal price p^* is

$$p^* = p^*(\gamma, \lambda_F) = \frac{1}{2}(\frac{\alpha}{k(1 - \gamma)} + c - \lambda_F).$$

Note that that dynamic prices are higher than static ones whenever $\lambda_F < 0$. The adjoint variable satisfies the equation

$$\dot{\lambda}_F = r\lambda_F + (p^* - c + \lambda_F)(\alpha - k(1 - \gamma)p^*)$$

and the terminal condition $\lambda_F(T) = 0$. Substitution from the maximum condition results in

$$\dot{\lambda}_F = r\lambda_F + (p^* - c + \lambda_F)^2 k(1 - \gamma)$$

which has the solution

$$\lambda_F(t) = -e^{rt} \int_t^T k(1 - \gamma)(p^* - c + \lambda_F)^2 e^{-r\tau} \, d\tau.$$

Hence we can conclude that $\lambda_F(t) \leq 0$. This is intuitive and is a consequence of the saturation effect.

The government's problem becomes

$$\max_{\gamma \in [0,1]} x(T)$$

subject to the constraints

$$
\begin{aligned}
\dot{x} &= (N - x)(\alpha - k(1 - \gamma)p^*) \\
p^* &= \frac{1}{2}(\frac{\alpha}{k(1 - \gamma)} + c - \lambda_F) \\
\dot{\lambda}_F &= r\lambda_F + (p^* - c + \lambda_F)(\alpha - k(1 - \gamma)p^*) \\
\lambda_F(T) &= 0.
\end{aligned}
$$

The Hamiltonian of the government is given by

$$H_G = \lambda_G(N - x)(\alpha - k(1 - \gamma)p^*) + \mu(p^* - c + \lambda_F)(\alpha - k(1 - \gamma)p^*) + \mu r\lambda_F,$$

where λ_G and μ are the adjoint variables which satisfy the equations

$$
\begin{aligned}
\dot{\lambda}_G &= \lambda_G(\alpha - k(1 - \gamma)p^*) \\
\dot{\mu} &= -\frac{1}{2}[(N - x)k(1 - \gamma)\lambda_G + \mu(\alpha - (c - \lambda_F)k(1 - \gamma))]
\end{aligned}
$$

Theorem 3 *In case of a linear price response function, the open-loop Stackelberg equilibrium for Model III consists of a maximal subsidy $\gamma^* \equiv 1$ and a maximum price set by the monopolist. Moreover, since $\mu(t) \equiv 0$, the open-loop Stackelberg equilibrium is time consistent and leads to the same qualitative results as the feedback equilibrium.*

5 Conclusions

This paper has analyzed the optimal subsidy policy for a government who wishes to accelerate the diffusion of a durable good by maximizing the cumulative number of units sold at a predetermined instant of time. The firm, knowing the subsidy policy of the government, sets the price of the good so as to maximize its discounted stream of profits. Using differential game theory we derive the optimal strategies as Stackelberg equilibria. It turned out that it is optimal for the government to pay maximum subsidies. This policy is a direct consequence of the government's objective function as well as the assumption that government does not face any budget constraint. In case the government's actions are restricted by a budget constraint, it still has the incentive to pay a subsidy that exhausts the available financial resources. This can, however, lead to a policy in which subsidies below the maximal level are granted (cf. Kalish and Lilien (1983), Zaccour (1995)). It will remain one of our future tasks to explore optimal subsidy scheems when the government is bounded by a limited budget.

6 References

Başar, T. and G. J. Olsder (1995). *Dynamic Noncooperative Game Theory* (2nd ed.). New York: Academic Press.

Dockner, E. J. and A. Gaunersdorfer (1995), "Strategic New Product Pricing When Demand Obeys Saturation Effects," forthcoming in *EJOR*.

―――― and S. Jørgensen (1988), "Optimal Pricing Strategies for New Products in Dynamic Oligopolies," *Marketing Science*, 7, 315–334.

Eliashberg, J. and A. P. Jeuland (1986), "The Impact of Competitive Entry in a Developing Market upon Dynamic Pricing Strategies," *Marketing Science*, 5, 20–36.

Kalish, S. (1983), "Monopolistic Pricing with Dynamic Demand and Production Cost," *Marketing Science*, 2, 135–159.

―――― and G. L. Lilien (1983), "Optimal Price Subsidy for Accelerating the Diffusion of Innovation," *Marketing Science*, 2, 407–420.

Zaccour, G. (1995), "A Differential Game Model for Optimal Price Subsidy of New Technologies," forthcoming in: *Game Theory and Applications*.

Optimal Pricing Strategies for Primary and Contingent Products under Duopoly Environment [1]

Yuko Minowa, Rutgers University, USA

S.Chan Choi, Rutgers University, USA

Abstract. The purpose of the current paper is to study optimal pricing policies for duopoly firms that introduce two new related products: a primary and a nondurable captive contingent product. The use of the former requires the latter product, and the diffusion and the sales of the latter are contingent upon the diffusion of the former. The competition is modelled as a differential game and the solutions are derived as open-loop Nash equilibria. The results of our analysis show that the prices of the contingency product are constant, determined by constant cost, shadow price and own- and cross-elasticities. On the other hand, the price trajectories of the primary product are characterized by such parameters as the directionality of the diffusion effect of the primary product, the magnitude of the coefficient of the repeat purchase of the contingent product and the convexity or concavity of the cumulative adoption of the primary product with respect to time.

1 Introduction

Companies in many industries do not merely produce and market a single product for their long-run survival. A multiple number of products that pass through stages of "cradle-to- grave" product life-cycle are frequently sold

---
[1]   We wish to thank Steffen Jørgensen, George Zaccour, Engelbert Dockner, Gary Erickson and Gila Fruchter who had provided us valuable comments on the previous version of the paper. The first author acknowledges the Research Fund from the Ph.D. Program in Management, Rutgers University. The second author acknowledges the Research Grant from the Faculty of Management, Rutgers University.

at the same time. Furthermore, the demand for one product at a certain time may be related to the demand for other(s) in some fashion. Such interdependence of demand for products implies the interdependence of pricing strategies as an immediate consequence in the following ways: The adoption of either product is partly dependent on the price and the performance (i.e., cumulative adoption) of the other. In turn, the pricing of either product is influenced by the pricing of the other.

Jeuland and Dolan (1982) point out that one of a major characteristics of the pricing environment that creates "complex mixing" of demand, supply and competitive consideration, is that a firm typically markets products that are interrelated on the demand and cost side. Therefore, they contend "it is suboptimal to determine the price of each product offering independently" (1982, p. 3). While the investigation of product line pricing and optimal bundle pricing have been popular subjects of research in marketing, the framework of the analyses is based on static model building. (The reader who is not familiar with dynamic pricing models in Marketing for monopoly and oligopoly environments is referred to Eliashberg and Chatterjee 1985, Dockner and Jørgensen 1988, Jørgensen 1986, Kalish 1983, 1988, Rao 1984, 1992, and the references contained therein.)

In recent years, some attention has been given to pricing strategies for multiple numbers of products in a dynamic context. The pricing problems in this area can be broadly classified into two categories based on the nature of the relationship between two (or more) products in question: (1) sequential replacement of one product, and (2) simultaneous diffusion of related products.

In the former, the problem is concerned with the pricing of two products sold in different time periods with or without overlap. Typically, the second product is an advanced version of the first product and usually appears in the market as a result of technological improvement. Examples of such a product relationship are found in such product categories as a black & white TV and a color TV, and a PC with a 80286 chip and a PC with a 80386 or 80486 chip. The dynamic pricing models for such products are developed by

Padmanabhan and Bass (1993) and Bayus (1993).

The latter type of pricing problem is concerned with the pricing of interdependent products in the same period of time. The relationship between two products may be characterized as substitutes, complements, or primary and contingent. While substitutes and complements are product relationships that have been studied in economics as well as in marketing, the third type has not gained much attention in either discipline. A contingent product (which is also called a secondary product) is such that its diffusion and sales are dependent upon the adoption and sales of another (primary) product. Contingent products are broadly classified into optional and captive contingent products. For the former, contingent product cannot be used without the primary product, while the primary product alone can function for its purpose. For example, a VCR cannot be used without a TV (while a TV, the primary product can be used without a VCR). As a consequence, the diffusion and sales of VCRs are, to large extent, dependent upon the diffusion of TVs. Captive contingent products are such that both the primary and contingent must be used together to function. The examples include camera and film, CD player and CD, photo copier and toner, etc. The diffusion for each of the above product relationships is modelled by Peterson and Mahajan (1978). Further, dynamic pricing strategies of a firm that introduce new primary and contingent products under a monopoly environment are studied by Mahajan and Muller (1991).

In the present paper, we derive optimal pricing strategies for duopoly firms that introduce both primary and nondurable captive contingent products. The remaining part of the paper is organized as follows. §2 develops the model and states our assumptions. §3 explains our analytical results regarding the optimal pricing policies for duopoly firms. Finally, §4 concludes our discussion with possible extensions of the current study and the managerial implications.

2 The Model

In the current scenario, there are only two firms in the market that are concerned with the pricing of both primary and contingent products. We call these two firms "integrated" duopoly firms. Let t denote time and define for firm i, $i = 1,2$:

$x_i = x_i(t)$ ... cumulative sales volume of firm i for primary product by time t,

$y_i = y_i(t)$ ... cumulative sales volume of firm i for contingent product by time t,

$p_{(x)i}$ ... price of primary product of firm i,

$p_{(y)i}$ ... price of contingent product of firm i,

$c_{(x)i}$ ... firm i 's constant marginal cost for primary product,

$c_{(y)i}$ ... firm i 's constant marginal cost for contingent product,

$x = x_1 + x_2$ ... cumulative industry sales of primary product by time t, and

$y = y_1 + y_2$ ... cumulative industry sales of contingent product by time t.

Then we specify dynamic demand functions $\dot{x}_i = dx_i/dt$ and $\dot{y}_i = dy_i/dt$ as multiplicative function of cumulative industry sales and price function:

$$\dot{x}_i(t) = f(x)\, h_i(p_{(x)1}, p_{(x)2}) \qquad (2.1.1a)$$

$$\dot{y}_i(t) = g(x)\, k_i(p_{(y)1}, p_{(y)2}) \qquad (2.1.1b)$$

$$i = 1, 2,$$

where $g(x) = \gamma x$ and γ is a coefficient of repeat purchase, which indicates the average usage rate for the captive product. The industry demand, $f(x)$, for the primary product, x, at time t can be expressed in a more specific manner, such as the Bass (1969) diffusion model. Since our intention is to build as general model as possible, however, we continue to denote the demand function by $f(x)$ while deriving the optimal pricing paths.

For the objective function, we assume that each integrated firm seeks to maximize the current value of its profit stream from the both primary and contingent products with respect to prices of both products (i.e., $p_{(x)i}$ and

$p_{(y)i}$) over the planning horizon:

$$\Pi^i = \int_0^T e^{-r_i t} \left[(p_{(x)i} - c_{(x)i})\dot{x}_i + (p_{(y)i} - c_{(y)i})\dot{y}_i \right] dt \qquad (2.1.2)$$

where r_i denotes a constant discount rate for firm i. In the following, we discuss the assumptions that underlie our model.

Dynamic demand equations. The unique aspect of the model is the dynamic demand function for the contingent product, (2.1.1b). The current demand for the primary product is a function of the cumulative sales of the product in the past and the current prices of two competing firms. On the other hand, the dynamic demand expression for the contingent product (2.1.1b) states that the demand for the product not only depends on the past sales and prices of the two firms, but also on the past sales of the primary product. More specifically, we assume that the ceiling (i.e., the total potential) of the market size for the contingent product at time t is the cumulative industry sales of the primary product x, while the ceiling for the primary product is some fixed number N.

The multiplicative demand expressions imply that there is a structure in aggregate consumption behavior in the market. These dynamic demand equations can be interpreted as follows: The demand of the brand of the product is first influenced by the non-price aspects of diffusion effect of the product (e.g., positive or negative word of effect, saturation effect). Those who decide to purchase a product are, in turn, influenced by the price of each firm to further decide the brand to purchase.

Now, let us define $\partial h_i / \partial p_{(x)i} = h_{i1}$, $\partial h_i / \partial p_{(x)j} = h_{i2}$, $\partial^2 h_i / \partial p_{(x)i}^2 = h_{i11}$, $\partial^2 h_i / \partial p_{(x)j}^2 = h_{i22}$, $\partial^2 h_i / \partial p_{(x)i} \partial p_{(x)j} = h_{i12}$, $\partial k_i / \partial p_{(y)i} = k_{i1}$, $\partial k_i / \partial p_{(y)j} = k_{i2}$, $\partial^2 k_i / \partial p_{(y)i}^2 = k_{i11}$, $\partial^2 k_i / \partial p_{(y)j}^2 = k_{i22}$, and $\partial^2 k_i / \partial p_{(y)i} \partial p_{(y)j} = k_{i12}$. Then assumptions made on the partial derivatives of h_i and k_i in the dynamic demand equations are: (1) $h_{i1} < 0$ and $k_{i1} < 0$ for own-price elasticity of demand, (2) $h_{i2} > 0$ and $k_{i2} > 0$ for cross-price elasticity of demand, and (3) $h_{i1} h_{j2} > h_{i2} h_{j1}$ and $k_{i1} k_{j2} > k_{i2} k_{j1}$, which signify that the magnitude of the change in demand for the firm's product is larger when caused directly by changing its own price than when caused indirectly by the action of its rival.

In addition, we assume a linear price function, i.e., $h_{i11} = h_{i22} = h_{i12} = 0$ and $k_{i11} = k_{i22} = k_{i12} = 0$. This assumption implies that the demand for the product is affected by the substitutability of the competing products resulting from the price differences, and there is no non-linear interaction between the prices.

Objective functions. Contrary to static analyses, duopoly firms maximize the current value of their profit stream over a finite time horizon, as noted in (2.1.2) above.

Constant marginal cost. Since the objective in this chapter is to focus on the strategic consequences of interacting demand dynamics of two products, we do not incorporate cost dynamics (i.e., an experience curve effect) into our model. Our analysis is performed for a finite time horizon. According to Eliashberg and Jeuland, for such an analysis, "the assumption of constant marginal cost is less severe" (1986, p. 23). We focus on the dynamic effect of cost on pricing decisions in the subsequent chapters.

Information structure. The model is deterministic and both duopoly firms know each other's demand and costs (i.e., complete information). We also assume open-loop information: The state information that firms utilize in establishing strategies is given at the beginning of the game and is not updated in subsequent time periods.

Equilibrium concept. In defining the criterion of optimality, the solution in our dynamic analyses uses the Nash solution where no collusive action is allowed between duopoly firms. In other words, the strategies chosen by two firms are optimal in the sense that neither party can obtain a better performance for itself while its rival continues to maintain its own Nash pricing strategy.

Now, let us establish the necessary conditions for an open- loop Nash equilibrium. Define the Hamiltonian function for each firm as:

$$H_i = (p_{(x)i} - c_{(x)1})\dot{x}_i + (p_{(y)i} - c_{(y)i})\dot{y}_i + \lambda_{ii}\dot{x}_i + \lambda_{ij}\dot{x}_j + \mu_{ii}\dot{y}_i + \mu_{ij}\dot{y}_j,$$

$$i = 1, 2, \quad j \neq i$$

where the adjoint variables λ_{ij} and μ_{ij} (i.e., shadow prices) satisfy the end-conditions $\lambda_{ij}(T) = 0$ and $\mu_{ij}(T) = 0$, supposing no salvage term is considered in our problems. The adjoint variables need to satisfy the system of differential equations $\dot{\lambda}_{ij} = -(\partial H_i/\partial x_j)$ and $\dot{\mu}_{ij} = -(\partial H_i/\partial y_j)$. Hence, the transversality conditions, $\lambda_{ij}(T) = 0$ and $\mu_{ij}(T) = 0$, along with the adjoint equations, $\dot{\lambda}_{ij} = -(\partial H_i/\partial x_j)$ and $\dot{\mu}_{ij} = -(\partial H_i/\partial y_j)$ imply $\lambda_{ii} = \lambda_{ij}$ and $\mu_{ii} = \mu_{ij}$. Therefore, we define $\lambda_i = \lambda_{ii} = \lambda_{ij}$, $\lambda_j = \lambda_{ji} = \lambda_{jj}$, $\mu_i = \mu_{ii} = \mu_{ij}$ and $\mu_j = \mu_{ji} = \mu_{jj}$. The optimal strategies are, therefore, the functions $p_i^*(t)$ that optimize the corresponding Hamiltonian functions H_i given λ_i, λ_j, μ_i, μ_j.

The above necessary condition with respect to adjoint variables can be then rewritten as:

$$\begin{aligned}
\dot{\lambda}_i &= -\frac{\partial H_i}{\partial x} \\
&= -(p_{(x)1} - c_{(x)i} + \lambda_i)f_x h_i - \lambda_i f_x h_j - (p_{(y)}i - c_{(y)i} + \mu_i)\gamma k_i \\
&\quad - \mu_i \gamma k_j \quad \text{and} \\
\dot{\mu}_i &= -\frac{\partial H_i}{\partial y} \\
&= -(p_{(y)i} - c_{(y)i} + \mu_i)g_y k_i - \mu_i g_y k_j.
\end{aligned}$$

$\dot{\mu}_i$ above equals to 0 since $g_y = 0$. For interior solutions (i.e., p_i positive), the maximum principles of firm i are:

$$\frac{\partial H_i}{\partial p_{(x)i}} = \dot{x}_i + (p_{(x)i} - c_{(x)i} + \lambda_i)\frac{\partial \dot{x}_i}{\partial p_{(x)i}} + \lambda_i \frac{\partial \dot{x}_j}{\partial p_{(x)i}}$$

$$\qquad + (p_{(y)i} - c_{(y)i} + \mu_i)\frac{\partial \dot{y}_i}{\partial p_{(x)i}} + \mu_i \frac{\partial \dot{y}_j}{\partial p_{(x)i}} = 0 \quad \text{and}$$

$$\frac{\partial H_i}{\partial p_{(y)i}} = \dot{y}_i + (p_{(y)i} - c_{(y)i} + \mu_i)\frac{\partial \dot{y}_i}{\partial p_{(y)i}} + \mu_i \frac{\partial \dot{y}_j}{\partial p_{(y)i}} = 0.$$

We can then rewrite these optimality conditions as follows:

$$p_{(x)i}^* = c_{(x)i} - \lambda_i - \frac{h_i}{h_{i1}} - \lambda_i \frac{h_{j1}}{h_{i1}} - \frac{(p_{(y)i} - c_{(y)i} + \mu_i)\gamma k_i x_{p_{(x)i}}}{fh_{i1}}$$

$$- \frac{\mu_i \gamma x_{p_{(x)1}} k_j}{fh_{i1}} \quad \text{and}$$

$$p_{(y)i}^* = c_{(y)i} - \mu_i - \frac{k_i}{k_{i1}} - \mu_i \frac{k_{j1}}{k_{i1}}.$$

Then, the dynamics of the prices are derived by differentiating the above solutions for the price with respect to time:

$$\dot{p}_{(x)i} = -\dot{\lambda}_i - \frac{d}{dt}\left(\frac{h_i}{h_{i1}}\right) - \frac{d}{dt}\left(\frac{\lambda_i h_{j1}}{h_{i1}}\right) - \frac{d}{dt}\left[\frac{(p_{(y)i} - c_{(y)i} + \mu_i)\gamma k_i x_{p_{(x)i}}}{fh_{i1}}\right]$$

$$- \frac{d}{dt}\left(\frac{\mu_i \gamma x_{p_{(x)1}} k_j}{fh_{i1}}\right)$$

$$= -\frac{2\dot{\lambda}_i h_{i1} h_{j2}}{4h_{i1} h_{j2}} + \frac{\dot{\lambda}_j h_{j2} h_{i2}}{4h_{i1} h_{j2}} + \frac{\dot{\lambda}_j h_{i2}^2}{4h_{i1} h_{j2}}$$

$$+ \frac{\dot{f}\gamma x_{p_{(x)j}} k_j h_{i2}(p_{(y)j} - c_{(y)j} + \mu_j + 1)}{f^2(4h_{i1} h_{j2})}$$

$$- \frac{2\dot{\lambda}_i h_{j2} h_{j1}}{4h_{i1} h_{j2}} - \frac{2\dot{f}\gamma x_{p_{(x)i}} k_i h_{j2}(p_{(y)i} - c_{(y)i} + \mu_i + 1)}{f^2(4h_{i1} h_{j2})}.$$

In the last equation, the optimal pricing path $\dot{p}_{(x)i}$ was derived by assuming $\dot{p}_{(y)i} = 0$. This assumption must hold since:

$$\dot{p}_{(y)i} = -\dot{\mu}_i - \frac{d}{dt}\left(\frac{k_i}{k_{i1}}\right) - \frac{d}{dt}\left(\frac{\mu_i k_{j1}}{k_{i1}}\right) = \frac{\dot{p}_{(y)i} k_{i1} k_{i2} k_{j1} k_{j2}}{4k_{i1}^2 k_{j2}^2}$$

when $\dot{\mu} = 0$. The last equation leads to either

$$\dot{p}_{(y)i} = 0 \quad \text{or} \quad \frac{k_{i2} k_{j1}}{4k_{i1} k_{j2}} = 1.$$

Since $k_{i1} k_{j2} > k_{i2} k_{j1}$ by our initial assumption, $\dot{p}_{(y)i}$ must equal to 0.

3 Optimal Pricing Policies

In the last section, we attempt to derive the optimal pricing paths by differentiating $p_{(x)i}^*$ and $p_{(y)i}^*$ with respect to time. In interpreting the results, we see that the directionality of the price paths is dependent mainly on the directionality of $\dot\lambda_i$ and $\dot f$, and concavity or convexity of the industry demand for the primary product.

The sign of $\dot\lambda_i$, in turn, depends partly on the positivity or negativity of the diffusion effect of the primary product, i.e., $f'(x) \lessgtr 0$. When there is a positive diffusion effect, the shadow price falls ($\dot\lambda_i < 0$). On the other hand, when there is a negative diffusion effect, the sign of $\dot\lambda_i$ is indeterminant; it is positive if 1) the shadow price λ_i is large ($\lambda_i \gg 0$); 2) the negative diffusion effect is large in magnitude ($| f'(x) | \gg 0$); 3) a differential effect of price function of the primary product is large and/or larger than that of the contingent product ($h_i \gg 0, h_j \gg 0$, and/or $h_i > k_i$ and $h_j > k_j$); and 4) the per unit profit of the primary product is larger than that of the contingent product ($(p_{(x)i} - c_{(x)i} + \lambda_i) > (p_{(y)i} - c(y)i + \mu_i)$).

On the contrary, when there is a negative diffusion effect, the sign of $\dot\lambda_i$ is negative if 1) the coefficient of repeat purchase is large ($\gamma \gg 1$); 2) the shadow price μ_i is large; 3) the differential effect of price function for contingent product is strong and/or stronger than that of the primary product ($k_i \gg 0, k_j \gg 0$, and/or $k_i > h_i$ and $k_j > h_j$); and 4) the per unit profit of the contingent product is larger than that of the primary product ($(p_{(y)i} - c_{(y)i} + \mu_i) > (p_{(x)i} - c(x)i + \lambda_i)$).

Thus, the four individual outcomes of $\dot p_{(x)i}$ from the combination of $\dot\lambda_i \gtrless 0$ and $\dot f \gtrless 0$ are analyzed along with the $\dot p_{(y)i}$, and summarized as follows.

Proposition 1
If duopoly firms have symmetric demand and objective functions, and if no discount rate or experience effects are considered, then $p_{(y)i}$, the price of the nondurable captive contingent product for each firm, is constant, determined by constant marginal cost, shadow price, and own-price as well as cross-price elasticities of demand.

Proposition 2

$$f'(x) > 0, \gamma \gg 1, \dot{f} > (<)0 \Rightarrow \dot{p}_{(x)i} < (>)0 \quad \forall t \in [0, T].$$

$$f'(x) > 0, \gamma \sim 1, \dot{f} > (<)0 \Rightarrow \dot{p}_{(x)i} > (<)0 \quad \forall t \in [0, T].$$

In other words, when there is a positive diffusion effect for the primary product (i.e., $f'(x) > 0$) and the repeat purchase rate of the contingent product is high (i.e., $\gamma \gg 1$), then $p_{(x)i}$ decreases for all t where $d^2x/dt^2 > 0$ and increases for all t where $d^2x/dt^2 < 0$ given that the planning horizon is long. On the contrary, if the repeat purchase rate of contingent product is small (i.e., $\gamma \sim 1$) the $p_{(x)i}$ increases for all t where $d^2x/dt^2 > 0$ and decreases for all t where $d^2x/dt^2 < 0$.

Proposition 3

When there is a negative diffusion effect for the primary product (i.e., $f'(x) < 0$) and the repeat purchase rate of the contingent product is high (i.e., $\gamma \gg 1$), then $p_{(x)i}$, the price of the primary product, decreases for all t where $\dot{f} > 0$ and increases where $\dot{f} < 0$. On the other hand, if the repeat purchase rate of the contingent product is not significant (i.e., $\gamma \sim 1$), then the price of the primary product decreases where $\dot{f} > 0$.

4 Conclusion and Future Research

An attempt was made to identify formerly unsolved problems in dynamic pricing models in marketing through a literature survey. Then, the monopoly dynamic pricing models for primary and durable as well as nondurable contingent products were extended to duopolies, with as general a demand equation as possible. When the contingent product is nondurable in nature and captive in relation to the primary product, the price of the contingent product approaches a constant as the cross-elasticity of the demand among any one of products is reduced to zero. Also, it seems that when the rate of sales for the captive contingent product is determined by the pre-specified proportion of the total industry demand for the primary product, then the shadow price does not vary throughout the finite planning horizon. Determinants of the price of the primary product, on the other hand, seem to include the strength of the effect of the sales of captive contingent product, the positivity or negativity of the diffusion effect of the primary product and the magnitude of the shadow price. These results indicate meaningfulness of the present study to marketing management as well as marketing science. The results show that pricing for a product should not be done in isolation when there is another product that is related in demand in a competitive market.

There are several possible ways to extend the current research. As an immediate extension of the problem, we could relax some of the assumptions used in solving the maximization problem by studying the problem numerically. For example, we assumed a linear price function in specifying the current demand, in partly for mathematical tractability. However, by numerically simulating the problem, we can cope with a nonlinear price function which my describe the real-world situation more accurately.

Another way to extend the study is to compare dynamic pricing strategies among various forms of competition. We have so far concentrated on deriving Nash equilibrium, by using the assumption of complete information. Duopoly firms however may not act as "rational," "fair," or "wise" as we expect. They may be quite myopic and determine price based solely on cost. They may cooperate in determining the price. Or, one of the firms may be a price leader, and the rival firm may be a follower. The optimal trajectory for these alternative competitive assumptions may be derived numerically.

Further, the current study is extended by developing a model for a closed-loop solution. While we have assumed that duopoly firms compete with each other with complete information from the beginning of their competition, in reality, the firms alter their pricing strategy by observing each other's performance and reviewing its own performance (e.g., by observing the market share) in each successive time period. However, it is well known that obtaining analytically a closed-loop solution is extremely difficult, if not impossible (Erickson 1990), even for a single product model. Thus, the study may be carried out numerically by simulating the problem.

Finally, we may compare the price levels and the paths of two firms under two alternative industrial structure: In certain industries, duopoly firms that manufacture and sell a primary product (e.g., CD players) may consider manufacturing and selling nondurable captive contingent goods for the primary product (e.g., CDs) by themselves. In other industries, the primary product and contingent product are manufactured by separate and unrelated firms (e.g., camera and film). At the same time, the markets for both primary and contingent products can be competitive (if there are two firms that manufacture and sell only primary products and two other firms that only manufacture and sell contingent products).

The first situation above is likely to occur when the technologies of manufacturing primary and secondary products are closely related. In such a situation, the cost of producing optional products may be merely marginal. In other cases, producing both products allows the manufacturing cost to drop faster in the long run than otherwise. On the other hand, in the second situation described in the previous paragraph, the converse can be true; the technologies of manufacturing primary and contingent products are not related and any substantial capital investment to develop and manufacture the related product may not justify the cost incurred to take such action. Since a firm with alternative courses of action should know whether such differences in market structure affects the firm's pricing strategies (i.e., price level as well as the price trajectory), we therefore should extend our model building analysis to investigate the effect of such alternative industrial structures on pricing.

References

Bass, Frank M. (1969), "A New Product Growth Model for Consumer Durables," *Management Science*, 15 (January), 215-27.

Bayus, Barry L. (1992), "The Dynamic Pricing of Next Generation Consumer

Durables," *Marketing Science*, 11 (Summer), 251-63.

Dockner, Engelbert and Steffen Jørgensen (1988), "Optimal Pricing Strategies for New Products in Dynamic Oligopolies," *Marketing Science*, 7 (Fall), 315-34.

Eliashberg, Jehoshua and Rabihar Chatterjee (1985), "Analytical Models of Competition with Implications for Marketing: Issues, Findings, and Outlook," *Journal of Marketing Research*, 22 (August), 237-61.

──────────── and Abel P. Jeuland (1986), "The Impact of Competitive Entry in a Developing Market upon Dynamic Pricing Strategies," *Marketing Science*, 5 (Winter), 20-36.

Jeuland, Abel P. and Robert J. Dolan (1982), "Aspect of New Product Planning: Dynamic Pricing," in *Marketing Planning Models, TIMS Studies in the Management Sciences, 18*, A.A. Xoltners, ed. New York: North-Holland Publishing Company, 1-21.

Jørgensen, Steffen (1986), "Optimal Dynamic Pricing in an Oligopolistic Market: A Survey," in *Dynamic Games and Applications in Economics*, T. Bassar ed., Springer Lecture Notes in Economics and Mathematical Systems, 265, 179-237.

Kalish, Shlomo (1983), "Monopolist Pricing with Dynamic Demand and Production Cost," *Marketing Science*, 2 (Spring), 135-59.

──────────── (1988), "Pricing New Products from Birth to Decline: An Expository Review," in *Issues in Pricing: Theory and Research*, T.M. Devinney ed., Lexington, MA: Lexington Books, 119-44.

Mahajan, Vijay and Eitan Muller (1991), "Pricing and Diffusion of Primary and Contingent Products," *Technological Forecasting and Social Change*, 39, 291-307.

Padmanabhan, V. and Frank M. Bass (1993), "Optimal Pricing of Successive Generations of Product Advances," *International Journal of Research in Marketing*, 10 (2), 185-207.

Peterson, Robert A. and Vijay Mahajan (1978), "Multi- Product Growth Models," in *Research in Marketing - Vol. 1*, J.N. Sheth ed., Greenwich, CT: JAI Press, Inc., 201-31.

Rao, Vithala R. (1984), "Pricing Research in Marketing: The State of The Art," *Journal of Business*, 57, S39-S59.

_____ (1993), "Pricing Models in Marketing," in *Handbook in Operations Research and Management Science, Vol 5.*, J. Eliashberg and G.L. Lilien, eds. Elsevier Science Publishers.

Impacts of Category Management and
Brand Management From a Retailer's Perspective

Subir Bandyopadhyay[1] and Suresh Divakar[2]

[1] Faculty of Management, McGill University, Montreal, Quebec, Canada H3A 1G5[1]
[2] College of Business Administration, SUNY at Buffalo, NY 14260, USA

Abstract. The marketing objective of a manufacturer is to maximize the profits of its brands. While a brand management strategy is geared towards maximizing the profit of a single brand, the category management policy tries to maximize the overall profitability of the entire product line.

Unlike a manufacturer who wants to maximize the profit of his own brand, a retailer is not interested in the profitability of any particular brand but concentrates on the overall category profit. In other words, a retailer acts a category manager with the objective of maximizing category profit. We argue that the category profit may be maximized when the retailer allows a few large manufacturers to practice brand management and the retailer manages the rest of the brands. We develop a game theoretic model for the three scenarios and show that the optimal category profit is indeed higher for the mixed strategy than the other two strategies.

Keywords. Brand Management, Category Management, Game Theory

---

[1] The first author gratefully acknowledges the financial support provided by the Faculty of Management Researach Grant, McGill University and the Social Sciences and the Humanities Research Council of Canada Grant #410-95-0732.

Introduction

In recent years, category management has emerged as a viable alternative to the more traditional brand management (Zenor 1994). While a brand management strategy is geared towards maximizing the profit of a single brand, the category management policy tries to maximize the overall profitability of the entire product line. This policy is particularly helpful to those companies that have multiple brands in a given category which are potential competitors of one another. In recent years, a few companies such as Procter & Gamble have successfully implemented category management practices for their brands. The academic literature has also taken note of this new management concept. For example, Zenor (1994) has demonstrated that for linear demand structure, category management of the entire product line may return higher profits than brand management. The reason for this, as Zenor claims, is that the category management policy improves coordination among company's brands and minimizes the brand cannibalization tendencies.

Category Management by the Retailer

Zenor (1994) deals with category management from a manufacturer's perspective. Category management by a retailer is a somewhat different concept. Unlike a manufacturer, a retailer is not interested in the profitability of any particular brand but concentrates on profitability across categories (Abraham and Lodish 1993). The retailer is interested in only carrying brands with high retail profit margin and high turnover rates. In other words, a retailer acts as a category manager with the objective of maximizing profit of the entire category.

The diverging profit objectives of the manufacturer and the retailer pose an interesting problem. A manufacturer will be interested to know if he is going to gain or lose under the retailer's category management as compared to his own brand management. It is quite possible that a manufacturer makes more profit under the retailer's category management policy than his own brand management policy. For example, a pouwerful retailer can act as an effective mediator in controlling prices, promotions and merchandising of competitors. Thus, a single manufacturer or brand cannot take unilateral decisions to dictate pricing and other policies for the category.

From the retailer's perspective also, two policies need to be studied carefully. Retailer's category management policy for all brands may not be always more profitable for the retailer as compared to the profit under individual brand management policies. The manufacturer of a dominant brand may have considerable expertise and knowledge about the market that he uses to set marketing policies. Under some circumstances, brand management policies of a number of brands may generate higher profit than that for the retailer's category management. For example, the aggressive marketing strategy of a major brand can generate primary demand for a category. As a result, the entire category benefits and the retailer's profit increases. Hence, it will be in the benefit of the retailer to let that brand set its policies freely.

An important point to consider here is the current power structure in the retail business. Big and sophisticated retailers wield considerable power against most of the manufacturers. They regularly influence pricing, feature advertising and merchandising policies for all brands thus sharing some of the responsibilities of the traditional brand manager. Moreover, the retailer has to set the marketing policies of his own private label brands. Thus, it is very important to know the impact of various policies a retailer can adopt such as category management for all brands or allow brand management by individual brands. Alternatively, a mixed strategy of category management for some brands and brand management for others should also be considered.

The mixed strategy is perhaps the most realistic in the present circumstances. Most retail product categories consist of a few dominant national brands. These brands are financially strong enough to create substantial brand equity by means of long term advertising campaign and satisfactory product performance. Thus, these brands may have strong power to "pull" consumers to the store to search for these brands. It is, therefore, in the interest of the retailer to stock these brands so that consumers are encouraged to visit the store.

Also, consumers often associate name brands with high quality (Rao and Monroe 1991, Zeithaml 1992). Moreover, consumers' perception of the quality of a store is influenced by the quality of the goods it sells. Hence, by stocking name brands, the retailer can send a high quality signal to the market. Under these circumstances, the retailer may allow considerable freedom to some dominant brands to set their own pricing and promotion policies.

The proposed study will investigate the profit impacts for both manufacturers and retailers under these three policies: category management by the retailer, independent brand management by individual brands and category management by the retailer for some brands and independent brand management for other brands. The rest of the paper is organized as follows: first, we present the three different decision structures analyzed as well as the basic demand model. Then, we deal with the profit maximizing conditions for all three decision structures. Using real UPC sales data, we then present empirical findings about optimal prices and profits for all brands in a given category. Finally, we discuss the results and their managerial implications.

Model

Decision Structure:

We assume that the market consists of one retailer and a number of manufacturers. Each manufacturer produces a single brand. The retailer is generally more powerful than most manufacturers. Store brands and private labels will fall under this category. In most cases, the retailer has the final the word in setting their marketing strategies. However, there may be a few dominant manufacturers who enjoy a lot of flexibility in setting their own marketing strategies. This is the case of part category management by the retailer and part brand management by some manufacturers. We believe that this power structure reflects the reality of retail business today when major retailers enjoy more power than most manufacturers. However, for the sake of comparison, we also consider the two extreme cases: the retailer tries to maximize overall category profit knowing the profit margin of all manufacturers - the case of category management by the retailer, and each manufacturer tried to maximize own profit knowing the retailer's margin - the case of brand management by the manufacturer.

In Table 1, we have summarized the above structures as well as other possible retail decision structures along with brief descriptions of each structure. As evident from the summary, strategies 2, 4, 5 and 6 have been addressed by Zenor (1994).

Table 1 A Summary of Decision Structures in the Retail Business Involving one Retailer and Multiple Manufacturers

Strategy	Previous Studies	Marketing Variables Included	Typical Scenario
1, Category Management by the Retailer	Choi (1991) (Theoretical model only, no empirical testing)	Only Price	The retailer has complete control on how the category is managed and also has perfect knowledge about the profit that each manufacturer makes. Most manufacturers are small compared to the large retailer.

2. Brand Management by all manufacturers	Zenor (1994)	Only Price	Manufacturers are more powerful than the retailer. The retailer relies on the marketing abilities of manufacturers and lets them set their own marketing policies. Manufacturers have perfect knowledge about the retailer's profit.
3. Part brand management by a few manufacturers and part category management by the retailer for the other brands	None		The retailer lets a few powerful manufacturers set their marketing policies but manages private labels and smaller brands. The retailer is more powerful than most manufacturers but less powerful than manufacturers of a few dominant brands.
4. Category management by all manufacturers	Zenor (1994)	Only Price	The retailer is small compared to all manufacturers who have multiple brands in the same category.
5. Brand management by some manufacturers, category management by others	Zenor (1994)	Only Price	A multi-brand manufacturer adopts a uniform strategy for all its brands, other manufacturers manage their brands individually.
6. Collusion by all	Zenor (1994)	Only Price	All manufacturers set a collusive marketing strategy so that profit is maximized for everyone.

In other words, Zenor (1994) has dealt with the issue of category management from the manufacturer's perspective. Our study addresses decision structures 1, 2 and 3. Thus, we address the issue of category management by the retailer and variations thereof.

Demand:

Similar to Choi (1991) and Zenor (1994), we assume a linear demand function. However, unlike previous models that included competitive pricing effects only, our model includes competitive merchandising effects also. Merchandising is an important decision strategy in retailing, (Bowman 1988) and hence its inclusion in the model makes it more realistic. The demand function is expressed as follows:

$$q_r = a_r + b_{ri}\, p_i + d_{ri}\, m_i \qquad (1)$$

where,
q_r = sales of brand r
a_r = a scaling parameter of brand r
b_{ri} = a sensitivity parameter for brand r sales to brand i price

p_i = retail price of brand i
m_i = merchandising level of brand i
d_{ri} = a sensitivity parameter for brand r sales to brand i merchandising level

In matrix form, the demand system can be represented as follows:

$$Q = A_0 + BP + DM \qquad (2)$$

where,
Q is a $R*1$ vector of brand sales
A_0 is a $R*1$ vector of scaling parameters
B is a $R*R$ matrix of sensitivities of row brand sales to column brand price

P is a $R*1$ vector of retail brand prices
D is a $R*R$ matrix of sensitivities of row brand sales to column brand merchandising levels
M is a $R*1$ vector of retail brand merchandising levels

Since our control variable is only price, we can combine the coefficient matrices A_0 and D for compactness, and call it A. Equation (2) then becomes:

$$Q = A + BP \qquad (3)$$

The brand profits are expressed as:

$$\Pi_{man} = (diag\ Q)\,'(W\text{-}C) \qquad (4)$$

where, C is a $R*1$ vector of production costs for each brand and W is a $R*1$ vector of wholesale prices which the retailer pay to the manufacturers.

Similarly, the retailer's profit can be expressed as follows:

$$\Pi_{ret} = Q\,'(P\text{-}W) \qquad (5)$$

where, W is a $R*1$ vector of wholesale prices which the retailer pay to the manufacturers.

First Order (Profit Maximizing) Conditions

In this section, we present the profit maximizing conditions for three decision structures. First, we present the solutions when all brands are under traditional brand management. Next, we deal with the solutions when the retailer acts as an active category manager for all brands. Finally, we present the solutions when the retailer manages most of the brands, but a few brands pursue traditional brand management.

Scenario 1:

Brand Management by All Manufacturers:

In this case, the retailer has less control over manufacturers and lets them set their own pricing and promotion strategies. Manufacturers know the retail price set by the retailer and use this information to maximize the wholesale prices. The objective function is, therefore, Equation 4 (manufacturers profit). The problem is to maximize Equation 5 (retailer's profit) with respect to the retail brand prices for given wholesale prices. Equation 4 is then maximized with respect to wholesale prices subject to the previously determined retail brand prices.

The equilibrium solution to this optimality problem is given by the following retail prices P^* and wholesale prices W^*:[2]

$$P^* = (B' + B)^{-1} (B'W - A)$$
$$W^* = [XB' + Y]^{-1} [XA + YC - A_0.]$$

Where,

$$X = B(B' + B)^{-i} \text{ and,}$$
$$Y = diag(XB')$$

---

[2] Detailed derivations are available from the first author.

Scenario 2:

Category Management by the Retailer for all Brands:

In this decision scenario, the retailer tries to maximize the category profit knowing the profit margin of each manufacturer. In effect, the problem is to maximize Equation 4 with respect to wholesale prices for given retail prices, and then maximize Equation 5 with respect to retail prices subject to the previously determined wholesale prices.

The equilibrium retail prices P^* and wholesale prices W^* for this optimality condition are given below:

$$W^* = [Diag(B)]^{-1}. \; [Diag(B). \; C - BP - A]$$
$$P^* = [B'K + K'B]^{-1} \; \{B'[Diag(B)]^{-1} \; [Diag(B). \; C - A]$$

$K'A\}$

where, $\qquad\qquad K = I + [diag(B)]^{-1}. \; B$

Scenario 3:

Category Management by the Retailer for Some Brands, Brand Management for Others:

This is perhaps the most real-life decision scenario. Here, the retailer knows the profit levels of most manufacturers but for a few dominant brands. These major brands often undergo independent ad campaigns to create brand loyalty. Since well known brand names are known to signal quality (Rao and Monroe 1989, Zeithaml 1988), it will be in the interest of the retailer to stock these brands.

For ease of exposition, we can partition all the matrices to write the demand systems for the brand and category management separately. The demand system for brands under brand management is expressed as:

$$Q_1 = A_1 + B_{11} P_1 + B_{12} P_2 \qquad\qquad (6)$$

and, for brands under category management, the demand system is:

$$Q_2 = A_2 + B_{21} P_1 + B_{22} P_2 \qquad\qquad (7)$$

The brand profits are expressed as:

$$\Pi_{man} = diag(Q_1') \; (W_1 - C_1) + diag(Q_2')(W_2 - C_2) \quad (8)$$

where suffices ?? 1 and 2 stand for brand management and category management respectively. All notations, e.g., A, B, P, W and C have the same meanings as in the first two scenarios.

The total retail profit is:

$$\Pi_{ret} = Q_1{}'(P_1 - W_1) + Q_2{}'(P_2 - W_2) \tag{9}$$

The optimal retail prices P_1 and P_2, and optimal wholesale prices W_1 and W_2 are given below:

$$P_1{}^* = \{B_{11}{}' K_1 + K_1{}' B_{11}\} \{B_{11}{}' [Diag (B_{11})]^{-1} [Diag (B_{11}) C_1 - B_{12} P_2 - A_2]$$

$$- K_1{}' A_1 - K_1{}' B_{12} P_2\} \tag{10a}$$

$$P_2{}^* = (B_{22} + B_{22}{}') [B_{12}{}^1 W_1 + B_{22}{}' W_2 - (B_{12}{}' + B_{21)P}1 - A_2] \tag{10b}$$

$$W_1{}^* = [Diag (B_{11})]^{-1} [Diag (B_{11}) C_1 - B_{11} P_1 - B_{12} P_2 - A_2] \tag{10c}$$

and, $\quad W_2{}^* = [Y_2 + Diag (Y_2)]^{-1}$

$$[X_2 A_2 + Diag (Y_2) C_2 - A_2 - X_{12} B_{12}{}' W_1 - \{B_{21} - X_2 (B_{12}{}' + B_{21})\}(P_2) \tag{10d}$$

where, $\quad X_2 = B_{22} (B_{22}{}' + B_{22})^{-1}$
$\qquad\qquad Y_2 = X_2 B_{22}{}',$ and
$\qquad\qquad K_1 = I + [Diag (B_{11})]^{-1} B_{11}$

Results and Discussion

We used 103 weeks of scanner data on ice-cream to empirically test our models. The data on three brands were made available by the Kroger Co. for the Cincinnati metropolitan area. One brand (BREYER'S) was a national brand and the other two (COUNTRY CLUB and TEXAS GOLD) were private label brands. Each brand comprised of several UPCs which varied only in terms of flavour and, therefore, were aggregated up to the brand level. Little information was lost in the aggregation process because all UPCs within the same brand had the same pricing and merchandising strategy.

Table 2 Demand Model Coefficient Estimates

Variable	Coefficient Estimate (standard errors in parentheses)		
	Country Club	Texas Gold	Breyers
Intercept	-3276.97 (3468.96)	67118 (9358.89)	13883 (2916.67)
Price-CClub	-760.1 (78.6)	1321.86 (2122.21)	-157.90 (661.38)
Price-Tx Gold	2641.85 (631.48)	-21877 (1703.66)	490.51 (530.94)
Price-Breyers	-702.86 (736.94)	661.97 (1988.19)	-3638.35 (619.62)
Merchandising-CClub	30.31 (4.72)	-0.132 (12.73)	2.35 (3.97)
Merchandising-Tx Gold	27.59 (6.38)	21.52 (17.22)	-2.99 (5.36)
Merchandising-Breyers	-6.40 (8.43)	-8.67 (22.75)	7.73 (3.09)
Summer	2125.45 (503.14)	1811.14 (1357.41)	4.54 (423.03)
Fall	2708.75 (498.19)	1885.45 (1344.01)	599.43 (418.88)
Spring	1287.85 (506.38)	747.92 (1366.16)	922.80 (425.76)
R^2	0.6098	0.7541	0.5755

Table 3 Comparison of Optimal Prices Under the Three Scenarios

Brand	Mean Price		Optimal Wholesale Prices			Optimal Retail Prices		
	W/S[1]	Retail	BM[2]	CM[3]	BM & CM[4]	BM	CM	BM & CM
Breyers	2.37	3.38	2.41	2.51	2.56	3.30	3.31	3.90
CClub	1.17	1.64	4.13	2.58	3.63	7.16	7.23	4.77
TX Gold	2.17	2.93	2.46	2.06	2.26	3.25	3.26	2.84

1. Wholesale price charge by the manufacturer to the retailer
2. Under the Brand Management scenario
3. Under the Category Management scenario
4. Under the Part Band Managment and Part Category Managment scenario

Table 4 Profitability Comparison[1]

Brand	Current Profits		New W/S Profits ('000)			New Retail Profits ('000)		
	W/S	Retail	BM	CM	Part BM & CM	BM	CM	BM & CM
Breyers	130	209	290	143	195	219	217	439
CClub	110	209	525	266	1225	972	956	916
TX Gold	469	494	781	396	875	1043	1040	925
Total	709	912	1596	805	2295	2234	2213	2280

1. Total profits for 103 weeks

In all cases the optimal profits are higher than the current profits. For Breyer's, the optimal wholesale price for the brand management (BM) scenario is about the same as the original price, but the optimal wholesale prices under category management (CM) and part brand - part category management scenario (BM + CM) are slightly higher. The profitability figures indicate that Breyer's is better off using the BM scenario where it makes the maximum profits of $290,000. However, for this brand, the retailer makes the most profit for the (BM + CM) case - more than twice the profit for the other cases. In fact, if we look at the combined profits of both the manufacturer and the retailer for this brand, the (BM + CM) scenario is the best with $634,000 (195K + 439K). Intuitively, this is the most realistic scenario with the manufacturer and the retailer pursuing their best objectives. This is a situation where the manufacturer can try to persuade the retailer to spend more on advertising or promote the Breyer's brand since the retailer makes much more profit than pure category management.

Looking at the results of the Country Club brand, it appears that it is under priced, both at the wholesale and retail level. The optimal prices are far higher than current prices and therefore the optimal wholesale and retail profits are also much larger. Here too, as with the Breyer's brand, the total wholesale and retail profits are the highest for the (BM + CM) case. For the Texas Gold brand, the retailer's profits are maximum for the CM scenario. Although this is the brand with the highest market share currently, its new profits are not very much higher than those for the other two brands. This may be because Texas Gold is losing market share to the other two brands.

Conclusions and Future Research

In this paper we developed a model that investigated three commonly used retailer - manufacturer strategies - (i) pure brand management of individual brands by the manufacturer, (ii) category management by the retailer and (iii) brand management for some brands and category management by the retailer for the remaining brands. We computed the optimal prices and profits for both the manufacturer and retailer under each of these scenarios and conclude that the optimal profits under the third scenario (part brand and category management) is highest.

The assumption of a liner demand is obviously not realistic and has been made for computational convenience. Future work will relax this assumption. Another possible avenue for future research is to incorporate merchandising variables also as control variables.

References

Abraham, Magid and Leonard Lodish (1993), "An Implemented System for Improving Promotion Productivity Using Store Scanner Data," *Marketing Science*, 12, No 3 (Summer) 248-269.

Bowmann, Russ (1988), "1987: The Year in Promotion," *Marketing and Media Decisions*, (July) 151-54.

Choi, S. Chan (1991), "Price Competition in a Channel Structure with a Common Retailer", *Marketing Science*, 10 (Fall) 271-96.

Rao, Akshay R. and Kent B. Monroe (1989), "The Effect of Price, Brand Name, and Store Name on Buyers' Perceptions of Product Quality: An Integrative Review", *Journal of Marketing Research*, 26 (August) 351-357.

Zeithaml, Valerie (1988), "Consumer Perceptions of Price, Quality, and Value: A Means-End Model and Synthesis of Evidence", *Journal of Marketing*, Vl 52 (July) 2-22.

Zenor, Michael J. (1994), "The Profit Benefits of Category Management", *Journal of Marketing Research*, 31 (May) 202-213.

Channel Coordination In The Presence of Two Sided Asymmetric Information

Ramarao Desiraju
University of Delaware, USA

Abstract. This paper develops a model of slotting allowances, taking into account the presence of two sided asymmetric information. The retailer's private information is the probability of success of the brand in the local market; the manufacturer's private information concerns the anticipated demand of its brand. In a brand-by-brand regime, some manufacturers are not given shelf-space because they are too likely to involve a low demand product; in the remaining cases, the retailer's slotting fee reveals the chance of success of the brand in the local market. The uniform method is found to increase ex ante profits when there is a larger proportion of high anticipated demand manufacturers in the retailer's market setting. The paper concludes with directions for further work.

1   Introduction

Effective coordination of channel activities has long been considered an important problem by both marketing practitioners and academics. In the last decade, several solutions to the coordination problem have been suggested by analytical researchers in marketing. The channel settings modeled by these researchers can be classified into three categories: The first category (e.g., Moorthy, 1987) deals primarily with channel settings characterized by a lack of information asymmetry or environmental uncertainty. The second category (e.g., Chu, 1992) studies scenarios in which the manufacturer is better informed than the retailer. The final category (e.g., Desiraju and Moorthy, 1995) models scenarios in which the retailer is better informed than the manufacturer. This paper extends the above body of research by studying channel settings in which there is two sided information asymmetry.

Specifically, we focus on the introduction of new products since in these introductions there exisits a potential for two sided asymmetric information. Each year, approximately ten to fifteen thousand new products are introduced in the grocery store business (e.g., The Chicago Tribune, 1988, Sullivan, 1989). In the face of such onslaught, a common problem faced by the retailer is to select only the good "quality" or high "demand" introductions. Since the manufacturer usually has better information (than the retailer) about the true innovativeness, quality, and demand of the introduction, the challenge facing the retailer is to sift out the bad quality introductions from the good ones.

On the other hand, the retailer[1] by virtue of her closeness, has better knowledge about the purchase patterns, demographics and other such information about her local customers. For example, say a manufacturer is introducing a new deodorant targeted for females over 55 years of age. Since the retailer is in a better position to know the age distribution of female shoppers in her store, she can better assess the chance of "high sales" or "success" for the new product. Since the information about age distribution is private knowledge for the retailer, she is better informed (than the manufacturer) of the likelihood of success of the introduction in the local market.

---

[1]Throughout the paper, we refer to the retailer as "she" and the manufacturer as "he."

Summarizing, even though the manufacturer knows the potential demand for his introduction, he cannot be completely certain how his brand will be received in the local market---since he does not have perfect access to local market information. Analogously, even though the retailer has local market information (e.g., she can judge the chance of success of an introduction in the product category), she cannot be completely certain about the demand (or quality) of the manufacturer's introduction. Since innovativeness of the product influences sales, the retailer cannot be certain of the sales that will be generated, even if the likelihood of acceptance by the local customers is high.

The above discussion highlights the source of the two-sided information asymmetry that we consider in our model. Our aim is to develop a coordinating mechanism or a contract between the manufacturer and the retailer that will resolve the asymmetry in an optimal manner, thereby improving the efficiency of the channel. In this paper, we focus on the retailer's use of slotting allowances[2] to achieve this goal. Our analysis reveals that in equilibrium, some manufacturers are not given shelf-space because they are too likely to involve a low quality product; in the remaining cases, the retailer's contract reveals the chance of success of the product category in the local market.

The rest of the paper is organized as follows. Section 2 develops a model from the perspective of a retailer who is dealing with a heterogeneous group of manufacturers and characterizes a sequential equilibrium (of the type suggested by Kreps and Wilson, 1982). Section 3 studies the effect of offering a uniform slotting allowance in the face of two sided information asymmetry.

---

[2]Note that Chu (1992) also studies the use of slotting allowances, but in the context of one sided information asymmetry.

Section 4 discusses the relative merits of the above two analyses and section 5 concludes the paper with directions for further work. All proofs are confined to an Appendix which is available from the author.

2   The Model

Let t = L, H denote the two types of manufacturers: low and high "anticipated demand" respectively. The manufacturer's type is assumed to be known only to the manufacturer. Assume that the chance of success of an arbitrary brand when sold through the retailer is represented by the probability $p \in [0, 1]$. This chance of success is a summary statistic of the extent and quality of the information which is available to the retailer[3], including any verifiable information obtained from the manufacturer (for instance, product quality certification or the results of market research conducted by an independent organization). Thus a manufacturer with a $p = 0.7$ is either high anticipated demand and unlucky to have a 30% chance of failure, or, low anticipated demand and lucky to have a 70% chance of success. I assume that the probability of success is private information possessed by the retailer. However, the manufacturer's type and the retailer's information are not unrelated; in particular, manufacturer's anticipated demand and chance of success are assumed to be jointly distributed. Let $G(p,t)$ denote the joint distribution of p and t: that is,

$G(p,t)$ = Prob{chance of success $\leq$ p and manufacturer is of type t}.

The expression which will be relevant to the retailer's decision is denoted $f(p)$ and represents the probability that, given the chance of success, the manufacturer is actually high anticipated demand:

---

[3]This is due to the emergence of "power retailers," who define their customer segments sharply, and have a good understanding of the needs and wants of those segments. It seems reasonable to assume that such a retailer has available to her a summary statistic such as the chance of success.

142

f(p) = Prob{ manufacturer is high anticipated demand| chance of success p}

$$= \frac{dG(p,H)}{dG(p,H)+dG(p,L)} \; ,$$

where dG(p,t) denotes the density with respect to p. The expression which will be relevant to the manufacturer's decision problem is denoted K(p|t) and represents the conditional distribution of p given t:

K(p|t) = Prob{chance of success ≤ p| manufacturer is of type t}

$$= \frac{G(p,t)}{G(1,t)} \; .$$

Although the process through which the manufacturer approaches the retailer may itself convey some additional information about the manufacturer's demand[4], this possibility is suppressed in the interest of expositional simplicity. For the present, it is assumed that the process of approaching the retailer is essentially random, with q denoting the proportion of high demand manufacturers among those who approach; that is, q = G(1,H). The distribution function G(p,t) is assumed to be common knowledge.

Assumption 1: *f'(p) > 0 for all p ∈ [0,1]; that is, the higher the chance of success, the greater is the likelihood of that manufacturer being of higher demand type.*

Define $E_t(p|\delta)$ to be the t-type manufacturer's expectation of p, given that p belongs to the set δ, where $\delta \subseteq [0,1]$.

Assumption 2: $E_H(p|\delta) \geq E_L(p|\delta)$ *for all δ; i.e., the distribution of (p,t) is such that, conditional on p ∈ δ, a high demand manufacturer faces a higher chance of success (in expectation) than does a low demand manufacturer.[5]*

---

[4]From expositional convenience, from now onwards, we'll use "demand" to mean "anticipated demand".

[5]See Appendix A for an example of G(.) that satisfies these assumptions.

Let S be the slotting fee demanded by the retailer, R the revenue[6] anticipated if the product is sold through the retailer; note that R is taken as exogenous by the retailer and the manufacturer[7]. One consequence of Assumption 2 is that, conditional on $p \in \delta$, a high demand manufacturer expects greater revenues from selling through the retailer--because this manufacturer is (on average) more likely to be successful--than its lower demand counterpart.

Let C_M denote the costs incurred by the manufacturer in distributing the product; I assume that both the low and high demand manufacturers have the same costs of distribution. Let C_R be the costs incurred by the retailer in carrying the product, and let γ denote the fraction of the revenues, R, that accrue to the retailer.[8]

Events are assumed to proceed in the following order. First, the manufacturer and the retailer observe their private information. The retailer, on the basis of its private information, demands a slotting fee in exchange for carrying the manufacturer's product. Then the manufacturer, on the basis of its private information and whatever inference it draws from the announced slotting fee, either accepts or rejects the offer. If the offer is accepted (i.e., if the manufacturer agrees to pay the required slotting fee) the retailer carries the product and both parties collect the associated payoffs. If the offer is rejected, each party makes its reservation level of profits from alternative business arrangements.

---

[6]It may be noted that an alternative specification such as $R_H > R_L$, where R_t represents the revenues from selling an t-type manufacturer's brand, also provides similar insights.

[7]In practice, though, the retailer (and/or the manufacturer) may be able to influence R through variations in marketing efforts, and this presents a direction for further work.

[8]At this point, an explicit bargaining stage could have been incorporated to determine the magnitude of γ; however, in order to focus on the optimal method of setting slotting fee, we assume that γ is fixed exogenously.

Given this sequence of moves, a *strategy* for the manufacturer of type t is a function $m_t(S)$ specifying the probability that the manufacturer accepts the slotting fee S. We can write the expected profits to a type-t manufacturer who is demanded a fee S and accepts it with probability ϕ as

$$\pi_t^M(S, \phi; \delta(S)) = \phi [E_t(p|\delta(S)) (1 - \gamma) R - S - c_M] + (1-\phi) \pi_0^M \qquad (1)$$

where $\delta(S)$ describes the manufacturer's *beliefs* upon observing S; that is the set of retailer types p which the manufacturer believes would ask for S; note that this is not subscripted because there is no reason for different manufacturers to have different conjectures about this set. However, the expectation is subscripted because the high and low demand manufacturers may assign different distributions over the set $\delta(S)$ because (p,t) are jointly distributed. When the announced slotting fee does not reveal p, then the two types of manufacturers may use different strategies, $m_t(S)$, for t = H,L.

However, if the slotting fee S reveals p, then the two types of manufacturers will have identical (degenerate) expectations about p given a fee S. In this case, they might as well use the same strategy m(S). This "symmetry" assumption is formalized below in Assumption 3.

Assumption 3: *When both manufacturer types are indifferent about accepting or rejecting a slotting fee S, they use the same strategy m(S).*

The objective function of a retailer involves two goals: avoid carrying the products of low demand manufacturers, and avoid not carrying the products of high demand manufacturers. Consequently, I assume the retailer gives a greater weightage or utility to carrying the products of high demand manufacturers. Therefore, when faced with its portion of the final revenues, γR, the retailer obtains a utility of $(\gamma + \lambda)R$, where $\lambda > 0$, when a high demand manufacturer's product is sold, and $(\gamma - \alpha)R$, where $\alpha > 0$, when a low demand manufacturer's product is sold; further, we assume that $\alpha > \gamma$ as this provides the most interesting case to investigate.[9]

A *strategy* for the retailer is a function $S(p)$ specifying the slotting fee demanded when the chance of success is p. Then expected retailer profits from a brand with a chance of success p when a slotting fee S is demanded can be written as:

$$\pi^R(p, S; m_H(S), m_L(S))$$
$$= f(p) \{ m_H(S) [(\gamma + \lambda)R + S - c_R] + (1 - m_H(S)) \, \pi_0^R \}$$
$$+ (1 - f(p)) \{ m_L(S) [(\gamma - \alpha)R + S - c_R] + (1 - m_L(S)) \, \pi_0^R \} \quad (2)$$

To interpret equation (2), assume that the retailer has observed a chance of success p and demanded a fee S. With probability $f(p)$ the manufacturer is of high demand, in which case he accepts the slotting fee with probability $m_H(S)$, and the retailer sells the product, incurring the carrying costs c_R but gaining the slotting fee and the utility of carrying a successful brand $(\gamma + \lambda)R$. With probability $(1 - m_H(S))$, the high demand manufacturer rejects the offer yielding the retailer its reservation profits from other available alternatives. With probability $(1 - f(p))$, the manufacturer is of low demand, in which case it will accept the slotting fee S with probability $m_L(S)$ forcing the retailer to carry the product--thereby, the retailer incurs a loss of $(\gamma - \alpha)R$ but gains the slotting allowance S, and incurs the costs of distribution c_R. With probability $(1 - m_L(S))$ the low demand manufacturer rejects the retailer's demand yielding the retailer's reservation profits.

When both manufacturer types use the same strategy $m(S)$, the above objective function simplifies to:

$$\pi^R(p, S; m(S)) = m(S) [a(p) R + S - c_R] + (1 - m(S)) \, \pi_0^R \quad (3),$$

---

[9]See footnote 5.

where $a(p) = \{(\gamma + \lambda)\, f(p) - (\alpha - \gamma)\,(1 - f(p))\}$. The expression $a(p)$ represents the expected net weight assigned to an additional unit of revenue from selling a manufacturer's brand, against which there is a chance of success p; the expectation arises from the fact that manufacturer's demand is unverifiable. Notice that $a'(p) = f'(p)\,(\alpha + \lambda) > 0$. Define $\psi(p) = a(p) + p(1 - \gamma)$, and select a p_0 such that $\psi(p_0) = \beta$, where $\beta = \dfrac{c_R + c_M + \pi_0^R + \pi_0^M}{R}$

Lemma 1: *If $p_0 \in (0,1)$ exists, it will be unique, and hereafter we assume its existence and interiority.*

Proof: Retailer will be indifferent between carrying and not carrying a manufacturer's product only when the following condition holds:

$$a(p)R + S - c_R = \pi_0^R.$$

In addition, the retailer wants to extract all the rents from the manufacturer. Consequently, the following condition also needs to hold:

$$p(1 - \gamma)R - S - c_M = \pi_0^M.$$

Solving the above two conditions for p and S, we obtain:

$\underline{S} = p_0(1 - \gamma)R - c_M - \pi_0^M$, and p_0 satisfies the condition:

$$\psi(p_0) = \frac{\pi_0^R + \pi_0^M + c_M + c_R}{R}, \text{ where } \psi(p) = a(p) + p(1 - \gamma). \qquad \text{Q.E.D.}$$

We use the following definition for a sequential equilibrium (for similar definitions see Kreps and Wilson, 1982, Cho and Kreps, 1987, or Reinganum, 1988).

Definition 1: A *sequential equilibrium* consists of beliefs $\delta^*(.)$ and strategies $(\,m_H^*(.),\, m_L^*(.),\, S^*(.)\,)$ such that

(a) $m_t^*(S)$ maximizes $\pi_t^M(S, \phi; \delta^*(S))$, $t = L, H$;

(b) $S^*(p)$ maximizes $\pi^R(p, S; m_H^*(S), m_L^*(S))$; and

(c) $\delta^*(S) \subseteq [0,1]$ for all S, and $\delta^*(S) = \{p | S = S^*(p)\}$ whenever this set is nonempty.

That is, the equilibrium strategy of each manufacturer type maximizes that manufacturer's expected profits, given the beliefs. The retailer's strategy maximizes the retailer's expected profits, given the anticipated responses of the two types of manufacturers. Finally, the manufacturers' beliefs are always confined to the set of retailer types which are known to be possible, and these beliefs are correct for equilibrium slotting allowances demanded.

Proposition 1: *A sequential equilibrium for this model is for the retailer to demand* $S^* = \infty$ *(i.e., not carry the product) if* $p \le p_0$; *otherwise demand* $S^* = p(1 - \gamma)R - c_M - \pi_0^M$.

Let $\underline{S} = p_0(1 - \gamma)R - c_M - \pi_0^M$, *and* $\bar{S} = (1 - \gamma)R - c_M - \pi_0^M$; *then, the manufacturer (whether high or low anticipated demand) accepts the offer.with probability* $\mathbf{m}^*(S) = 0$ *if* $S > \bar{S}$, *with probability* $\mathbf{m}^*(S) = 1$ *if* $S < \underline{S}$,

and with probability
$$\mathbf{m}^*(S) = \exp\left[\int \frac{-dS}{\psi\left(\frac{S + c_M + \pi_0^M}{(1-\gamma)R}\right) - \psi\left(\frac{\underline{S} + c_M + \pi_0^M}{(1-\gamma)R}\right)}\right]$$

for $S \in [\underline{S}, \bar{S}]$ *and with. Note that* $\mathbf{m}^*(S) = 0$ *at* $S = \underline{S}$.

Finally, the manufacturers' beliefs are $\delta^*(S) \in [0, p_0)$ *for* $S = \infty$, $\delta^*(S) = p_0$ *for* $S \in (\bar{S}, \infty)$, $\delta^*(S) = \frac{S + c_M + \pi_0^M}{(1-\gamma)R}$ *for* $S \in [\underline{S}, \bar{S}]$, *and* $\delta^*(S) = 1$ *for* $S < \underline{S}$.

Notice that sufficiently weak brands--those with $p \le p_0$--are not carried. How weak is "sufficiently weak" depends on $\psi(p)$ and β, which take into account the form of the inference function $f(.)$, the benefits (losses) accrued to a retailer from carrying high (low) demand manufacturers, on γ-- which determines how revenues are split between the manufacturer and the retailer--and on the costs of distribution, the anticipated revenues and the reservation profits. It is interesting that the relative weights given by the retailer to carrying high anticipated demand brands and low anticipated demand brands, and the profits from alternative arrangements (for instance, the profits from carrying another brand) influence whether the retailer will refuse to carry some brands. Overall, the results conform to the often quoted maxim: "Most retailers believe they should reject the brand unless they are convinced of the manufacturer's demand."

Formally, the equilibrium consists of two portions. One involves complete pooling for retailer types $p < p_0$ while the other involves complete separation for types $p \geq p_0$. Under the assumption that when the demanded slotting fee reveals the chance of success both types of manufacturers use the same strategy $m(S)$, the uniqueness proof of Reinganum and Wilde (1986) can be adapted to show that the separating portion of the equilibrium is unique. If this symmetry assumption is relaxed, I have been unable to rule out the possibility of a separating equilibrium in which the two manufacturer types use different equilibrium strategies.

Another expression of interest is the equilibrium probability of carrying the product as a function of its chance of success. This is the composition of the equilibrium probability of acceptance and the equilibrium demand of slotting fee: $m^{**}(p) = m^{*}(S^{*}(p))$, or,

$$m^{**}(p) = \exp\left[((1 - \gamma)R) \left(\int \frac{-dp}{\psi(p) - \psi(p_0)} \right) \right].$$

Proposition 2: *(a)   For $S \in (\underline{S}, \bar{S})$, the equilibrium probability of acceptance $m^{*}(S)$ decreases with an increase in the fee demanded S, and with an increase in the manufacturer's cost of distribution,* c_M.

(b) For p ∈ [p_0,1], the equilibrium slotting fee $S^*(p)$ demanded in exchange for carrying the product decreases with the manufacturer's cost of distribution, C_M, the portion of revenues that accrue to the retailer, γ, and the manufacturer's reservation profits, π_0^M; and increases with the chance of success, p, of the brand, and the revenue, R, anticipated upon carrying the brand. It is unaffected by the parameters C_R and π_0^R.

(c) For p ∈ (p_0,1], a manufacturer's equilibrium probability of accepting the demanded slotting allowance, $m^{**}(p)$, increases with the chance of success p, and the revenue, R, anticipated upon selling through the retailer; $m^{**}(p)$ decreases with the portion of revenues that accrue to the retailer, γ.

It should be noted that the effects of p and π_0^M upon S^* are in opposite directions; while p increases the slotting allowance, the manufacturer's alternative distribution arrangements tend to reduce the slotting allowance. An important implication of this equilibrium is that when a brand is not rejected by the retailer, the likelihood that it will be resolved by a rejection by the manufacturer to pay the slotting allowance is greater the smaller the margin obtained by the manufacturer. One potential direction for further research is to check whether the comparative static implications of this model are also consistent with the empirical reality of slotting allowances.

3   Restricted Discretion

The model in the previous section involved considerable discretion upon the part of the retailer, and in this section we investigate whether such discretion is desirable. Since the retailer cannot accurately anticipate the actual demand of a manufacturer's product, it appears inequitable to impose different slotting fees on manufacturers that are otherwise selling similar products. Such inequity can be eliminated if the retailer confines itself to demand a uniform slotting allowance from all manufacturers in that product category. This section examines whether such a restriction on the retailer's discretion can enhance its *ex ante* expected profits. In order to address that question, it is first necessary to characterize the equilibrium behavior when the uniform method is used.

When the retailer is required to demand the same fee from every manufacturer who wants to sell in a product category, independent of the chance of success of a particular brand, it must make a "pooling" offer. This can result in self-selection by the high and low demand manufacturers, because they have different ex ante expected values of p. Let E_t denote the t-type manufacturer's prior expectation over p:

$$E_t = \int_0^1 p \ dK(p|t)$$

By Assumption 2, $E_H \geq E_L$; the high demand manufacturer expects a greater likelihood of success than does a low demand manufacturer. If a pooled offer S is made, it is the expectation E_t which governs the manufacturer's decision. The t-type manufacturer's profits are:

$$\pi_t^M(S, \phi;[0,1]) = \phi[\ E_t(1 - \gamma)R - S - \ c_M] + (1 - \phi) \ \pi_0^M$$

The low demand manufacturer rejects S if and only if $S > S_0$, where

$$S_0 = E_L(1 - \gamma)R - c_M - \pi_0^M,$$

while the high demand manufacturer rejects S iff $S > S^0$, where

$$S^0 = E_H(1 - \gamma)R - c_M - \pi_0^M.$$

Thus any slotting fee $S \in [0, S_0]$ will be accepted by both types of manufacturers; any slotting fee $S \in (S_0, S^0]$ will be rejected by the low demand manufacturer and accepted by the high demand manufacturer; and, any slotting fee $S \in (S^0, \infty]$ will be rejected by both manufacturer types. The

retailer must determine an optimal fee, given the above anticipated behavior
of the manufacturer types.

If a demand $S \in (S_0, S^0]$ is made, it is accepted by the high demand and
rejected by the low demand manufacturers; thus the best such offer is $S = S^0$,
which yields *ex ante* expected retailer profits:

$$\pi_1^R = q [\ (\gamma + \lambda)R + S^0 - C_R] + (1 - q) \ \pi_0^R \tag{5}$$

Any offer $S > S^0$ is rejected by all manufacturers, yielding *ex ante* expected
retailer profits

$$\pi_2^R = \pi_0^R \tag{6}$$

Finally, an offer $S \leq S_0$ is accepted by all manufacturers; the best such offer is
$S = S_0$, yielding *ex ante* expected profits

$$\pi_3^R (S) = q [(\gamma + \lambda)R + S_0 - C_R] + (1 - q) [(\gamma - \alpha)R + S - C_R]$$
$$= \left[q (\alpha + \lambda) - (\alpha - \gamma) \right] R + S_0 - C_R . \tag{7}$$

A comparison of equations (5), (6), and (7) yields the following
characterization of the equilibrium with restricted discretion.

Proposition 3: *Define two intervals:*
$$I_1 = [0, \frac{E_L (1 - \gamma) - (\beta + \alpha - \gamma)}{E_H (1 - \gamma) - (\beta + \alpha - \gamma)}],$$

$$I_2 = (\frac{E_L (1 - \gamma) - (\beta + \alpha - \gamma)}{E_H (1 - \gamma) - (\beta + \alpha - \gamma)}, 1],$$

where, $\beta = \dfrac{C_R + C_M + \pi_0^R + \pi_0^M}{R}$.

When $q \in I_1$, *the optimal slotting fee under the uniform method is* $S = S_0$; *that is, both types of manufacturers' products are carried when the stipulated slotting allowance is paid. When* $q \in I_2$, *the optimal slotting fee under the uniform method is* $S = S^0$; *that is high demand manufacturers pay the allowance and obtain shelf space while the low demand manufacturers are not carried by the retailer.*

Expected ex ante retailer profits with restricted discretion are given by

$$E\,\pi^{R\,r} = \begin{cases} [\,q(\alpha + \lambda) - (\alpha - \gamma)\,]R + S_0 - c_R & q \in I_1 \\ q[(\gamma + \lambda)R + S^0 - c_R] + (1 - q)\pi_0^R & q \in I_2 \end{cases}.$$

It is interesting to note that when the retailer types separate (that is, the slotting fee is a revealing one) as developed in Section 2, the manufacturer types pool; but when the retailer types pool, the manufacturer types may separate (i.e., the low demand types reject S^0 and the high demand types accept the fee). The nice feature of such a screening outcome is that the retailer knows that each brand it carries is a high demand manufacturer.

4 The Preferred Regime

In this section, we wish to compare equilibrium *ex ante* expected retailer profits under the brand-by-brand and the uniform regimes described in Sections 2 and 3, respectively. A convenient characterization of the circumstances under which the brand-by-brand regime is preferred can be made in terms of the exogenous expression q.

Define

$$\pi^{R*}(p) = \pi(p, S^*(p); m^*(S^*(p))).$$
$$= m^*(S^*(p))\,[\,\psi(p)R - c_M - c_R - \pi_0^M - \pi_0^R\,] + \pi_0^R$$

By the envelope theorem,
$$\frac{\partial \pi^{R*}(p)}{\partial p} = \frac{\partial}{\partial p}\pi(p, S^*(p); m^*(S^*(p)))$$
$$= m^*(S^*(p))\left(\frac{\partial \psi(p)}{\partial p}R\right) > 0, \text{ for } p > p_0.$$

Thus, $\pi^{R*}(p) > \pi_0^R$, for $p > p_0$, while $\pi^{R*}(p) = \pi_0^R$ for $p \le p_0$. The density function for p can be written as: $q\,dK(p|H) + (1 - q)\,dK(p|L)$. The retailer's equilibrium ex ante expected profits with the brand-by-brand method can be written:

$$E \pi^{R*} = \int_0^1 \pi^{R*}(p) \, q \, dK(p|H) + \int_0^1 \pi^{R*}(p)(1-q) \, dK(p|L).$$

The expression $E \pi^{R*}$ can be compared with ex ante expected retailer profits under the uniform method, $E \pi^{R\,r}$, to obtain the following result.

Proposition 4: *For given values of the parameters E_H, E_L, γ, λ, α, c_R, c_M, R, π_0^M, and π_0^R, there exists a unique $q_0 \in (0, 1)$ such that the brand-by-brand method is preferred for $q < q_0$ and the uniform method is preferred for $q > q_0$.*

In the regime of the brand-by-brand method the retailer can selectively reject brands while retaining the ability to extract rents from the brands that are given shelf-space. However, under this method there is always a fraction of brands involving high anticipated demand manufacturers which are not carried. These two factors are traded-off against each other in selecting the optimal method of setting slotting allowances. Intuitively, when the process of approaching the retailer does not do a good job of screening out the low demand manufacturers (i.e., $q < q_0$), then unrestricted retailer discretion--i.e., the brand-by-brand method--is preferred. On the other hand, when the process of approaching the retailer does a good job of eliminating the low demand manufacturers (i.e., $q > q_0$), the restriction to a uniform offer is preferred.

With a sufficiently large proportion of high demand manufacturers, the preferred regime is the one with restricted discretion--since the retailer cannot afford to not carry a high demand manufacturer. On the other hand, when the proportion of high demand manufacturers is not sufficiently large, it becomes essential for the retailer to be very selective--i.e., the costs of carrying a product with a low chance of success is quite high, and the retailer prefers the brand-by-brand method, even though in this process, some of the high demand manufacturers may reject the slotting fee offer.

We can determine which discretionary regime is preferred by each type of manufacturer by comparing the expected profits made when selling through the retailer. So, the relevant expression from the brand-by-brand regime is, $E\pi_t^M = \int_{P_0}^{1} (p\gamma R - S - c_M) dK(p|t)$; this is compared with the expected profits from the uniform regime. For the L-type manufacturer, expected profits <u>through the retailer</u> are:

$$E\pi_L^M = \begin{cases} \pi_0^M & q \in I_1 \\ 0 & q \in I_2 \end{cases}.$$

For the H-type manufacturer, the expected profits <u>through the retailer</u> are:

$$E\pi_H^M = \begin{cases} \pi_0^M + \gamma R\left(\int_0^1 p\,dK(p|H) - \int_0^1 p\,dK(p|L)\right) & q \in I_1 \\ \pi_0^M & q \in I_2 \end{cases}$$

where the sets I_1 and I_2 are those defined in Proposition 3. For $q \in I_1$ or I_2, the high anticipated demand manufacturer prefers the uniform method; for $q \in I_1$, low anticipated demand manufacturer prefers the uniform method; and for $q \in I_2$, low anticipated demand manufacturer prefers the brand-by-brand method. These results can be combined with those of Proposition 4 to show that each of the following preference patterns arises from some configuration of parameters.

Proposition 5: *1. Suppose $q_0 \in I_1$. (1a) If $q < q_0$, then both types of manufacturers prefer the uniform method, while the retailer prefers the brand-by-brand method; (1b) If $q > q_0$, and $q \in I_1$, then all parties* *unanimously prefer the uniform method; (1c) If $q > q_0$, and $q \in I_2$, then the retailer and the H-type manufacturers prefer the uniform method, while the L-types prefer the brand-by-brand method.*

2. Suppose $q_0 \in I_2$. (2a) If $q < q_0$, and $q \in I_1$, then both types of manufacturers prefer the uniform method, while the retailer prefers the brand-by-brand method; (2b) If $q < q_0$, and $q \in I_2$, then the retailer and the L-type manufacturers prefer the brand-by-brand method, while the H-types prefer the uniform method; (2c) If $q > q_0$, then the retailer and the H-type manufacturers prefer the uniform method, while the L-type prefers the brand-by-brand method.

For most parameter values, there will be some disagreement among the interested parties regarding the preferred way of setting slotting allowances; so the existence of controversy and conflict (with respect to slotting allowances) among retailers and their vendors is not surprising. When unanimity does occur, it favors the regime of restricted choice--i.e., the uniform way of setting the allowance is preferred.

5 Conclusion

This paper provides an analysis of a market setting with two sided-asymmetric information. In addition, it extends the current research on slotting allowances. The analysis of the models developed in the paper identified some conditions under which the brand-by-brand method is preferred over the uniform method and vice versa.

The essential features of slotting allowance equilibrium which emerged from our analysis are that under a brand-by-brand scheme, sufficiently weak brands are rejected outright. A uniform-offer restriction upon retailer discretion was found to improve *ex ante* expected profits when the proportion of high demand manufacturers is sufficiently large. Typically, the retailer prefers to carry high demand manufacturer products under the uniform method and prefers to carry both high and low demand manufacturers under the brand-by-brand method; this is because sufficiently large slotting allowances are taken from the low demand type manufacturers. When the proportion of high demand manufacturers among those who approach the retailer is low, the benefits from the brand-by-brand method outweigh the costs of carrying low demand type manufacturers.

These results have been obtained in a very stream lined model--the only controllable variable was the slotting allowance. The next step in understanding these methods of setting slotting allowances is to consider other factors that are in the contract between manufacturers and their retailers. For instance, gross margins and amounts of cooperative advertising need to be studied along with slotting allowances. While the basic tradeoff identified in this paper will continue to hold in the more complex contract, these effects are likely to interact with other features. Carefully developed models are necessary to understand these interactions.

Another area is to work on is the problem of multiple equilibria. It will be useful to apply specific refinement concepts that will imply particular restrictions on out-of-equilibrium beliefs. Due to the difficulty of applying these concepts to the case of two-sided asymmetric information, this has been left for further research. A final area of further research is to provide an empirical test of the implications generated from this paper and from previous research, such as Chu (1992).

6   References

Advertising Age (1987),"Kodak Brand Calls Retreat in the Battery War," October 15, 3, 69.

_____ (1990),"Grocer 'Fee' Hampers New Product Launches," August 3, 1.

Banks, Jeffrey and Joel Sobel (1987),"Equilibrium Selection in Signaling Games," *Econometrica*, 55 (May), 647-662.

Cho, In-Koo and David Kreps (1987),"Signaling Games and Stable Equilibria," *Quarterly Journal of Economics*, 102 (May), 179-222.

Chu, Wujin (1992),"Demand Signaling and Screening in Channels of Distribution," *Marketing Science*, 11 (3), Summer, 1-21.

Desiraju, Ramarao and K. Sridhar Moorthy (1995),"Channel Coordination With Retailer Performance Requirements," *Working Paper*

Kreps, David and Robert Wilson (1982),"Sequential Equilibria," *Econometrica*, 50 (July), 863-894.

Moorthy, Sridhar (1987),"Managing Channel Profits: Comment," *Marketing Science*, 6(Fall), 375-379.

Reinganum, Jennifer and Louis Wilde (1986),"Settlement Litigation and the Allocation of Litigation Costs," *Rand Journal of Economics*, 17 (Winter), 557-566.

_____ (1988),"Plea Bargaining and Prosecutorial Discretion," *American Economic Review*, 78 (4), September, 713-728.

Sullivan, Mary (1989),"Slotting Allowances: An Inquiry," Mimeo, University of Chicago, Chicago, IL.

The Wall Street Journal (1988),"Supermarkets Demand Food Firms' Payments Just to Get on the Shelf," November 1, section A, 1, 9.

DYNAMIC MARKETING STRATEGIES IN A TWO-MEMBER CHANNEL [1]

Steffen Jørgensen and Georges Zaccour

Department of Management, Odense University, Denmark

GERAD, École des Hautes Études Commerciales, Montréal, Canada

Abstract

The paper is concerned with intertemporal conflict and cooperation between a manufacturer and a retailer in a vertical marketing channel. The setup is a differential game model of pricing and advertising. First, we identify a number of issues that must be addressed when the parties wish to establish and sustain a coordinated outcome. Second, we study Markovian marketing strategies in a game of simultaneous play. These outcomes will serve as disagreement outcomes if cooperation fails to be established. Third, a cooperative solution is identified, invoking some basic requirements of such a solution.

Key words: Channel dynamics, coordination, pricing, advertising

1   Introduction

The paper studies a dynamic channel relationship between a single manufacturer and a single retailer, using a differential game model. Application of dynamic game theory is not usual in the marketing channel literature (see, however, Chintagunta and Jain (1992)). Predominantly, the literature has employed static games that,

---

[1]Research supported by NSF, Denmark and NSERC, Canada

profit than what the firm would obtain in case of no agreement. In the negotiations to reach an agreement, the noncooperative profits act as the firms' reservation/conflict/disagreement payoffs[3]. Side payments between the firms may be needed; any such transfer payment scheme must satisfy individual rationality. Note that lump sum transfers can be unsatisfactory for the paying agent unless the agreement is binding. Continuous transfers can be made contingent upon whether the receiving agent has acted in accordance with the agreement under which the transfer is paid. To define individual rationality in dynamic games, the notion of agreeability has been suggested. Denote the state of the game at time $t \in [0, T]$ by $G(t)$. The prolongation of a strategy is the part of an overall strategy on $[0, T]$ that applies from an arbitrary $(t, G(t))$ till the terminal instant T. An agreement is agreeable at the initial point $(0, G_0)$ if, for both players, the agreement's prolongation at any $(t, G(t))$ along the agreed path is Pareto-efficient and payoff-dominates the prolongation of the disagreement strategy.

Group rationality is taken to mean that an agreement must be Pareto-efficient. Thus, if one firm wants to increase its profit along the efficient frontier, it has to be at the expense of the other firm. Most often each firm's noncooperative payoff falls short of what the firm could obtain in an efficient solution. This provides an incentive to cooperate.

The set of agreements satisfying individual and group rationality is not a singleton. This leads to the problem of how to select the particular agreement to be implemented: a bargaining (negotiation) problem arises. To determine a specific agreement, various procedures have been suggested, among those the axiomatic schemes that codify certain basic principles as axioms. The process by which the final outcome is selected is not modelled. The most prominent axiomatic scheme is the Nash scheme. Under certain regularity conditions, it selects a unique solution which is individual

---

[3] The possibility of ending up by receiving no more than these profits is a threat that should discourage the firms not to obtain an agreement.

rational and Pareto optimal. Sometimes an explicit bargaining scheme is not invoked; instead one looks for a coordinated solution satisfying individual rationality and efficiency.

Even if an agreement satisfying the above requirements can be identified, nothing guarantees that the agreement is enforceable. It is not certain whether a player will feel tempted to cheat, that is, to deviate unilaterally from his part of the agreement. Enforceability is, however, guaranteed if the parties have committed to implement their parts of the deal, or a binding contract has been signed[4]. From the theory of repeated games we know that cooperative equilibria can be strategically supported as equilibria (they become self-enforceable) if players are sufficiently patient and adopt strategies that threaten, if cheating is detected, by reversion to sufficiently grim punishment strategies. In this sense, self-enforcability means that no player can benefit from deviating unilaterally from the agreement, under the provision that current compliance with the agreement is followed by future compliance, and the threats to punish current defection are effective and credible.

Self-enforcing agreements have been intensively studied in game theoretical oligopoly literature, typically to explain the occurrence of tacit cooperation between oligopolists who neither communicate nor negotiate. The approach has also been applied in differential games, in particular in games of natural resource harvesting. In marketing channel coordination, however, the setting is different from oligopolistic competition and rivalry in natural resource harvesting. It seems less likely that a manufacturer and a retailer are antagonistic opponents of the kinds we observe in oligopolistic battles for market shares or in the exploitation of scarce natural

---

[4] Corfman and Gupta (1993) note that it is possible that the players would agree on an outcome that is not necessarily self-enforcing. Nevertheless, such an agreement could be enforced from other reasons: external factors (public opinion) as well as internal factors (prestige, credibility, ethics). The importance of such factors has been emphasized in marketing channel literature (e.g., Johansson and Mattsson (1988)), but little is said about the strategic mechanisms that make such relationships sustainable.

resources. The threat approach to enforceability seems less appropriate in problems of channel coordination.

Is the agreement going to cover only a part of, or the entire planning horizon? This raises important issues of renegotiation and commitment. Sometimes a single, long-term agreement is negotiated once and for all; in other cases it is possible to reopen negotiations at certain future instants of time (see, e.g., Rey and Salanie (1990)). Consider a bargaining situation in which one part is in a position to propose an agreement on which the other part can act only by acceptance or rejection. Generally, it is in the proposing part's best interest to suggest a long-term contract and such a commitment makes the proposing part better off. On the other hand, none of the firms may find it advantageous to commit any further than to a one-period contract[5]. The length of a subinterval could be interpreted to mean the minimum length of time of commitment to an agreement. In a channel context it does not make very much sense to suppose that there is no commitment at all, in the sense of the repeated game threat story. Recontracting could mean that at any instant of negotiation, a completely new contract, to cover the rest of the game, is bargained, or a spot contract, to cover the time interval until the next instant of renegotiation, is bargained. Notice that, from the point of view of the proposing agent, an optimal long-term contract cannot in general be implemented through a sequence of spot contracts. The reason is that the latter do no allow for intertemporal transfers of payoff[6]. Finally note that agents may have the possibility to renegotiate but it could be in everyone's best interest if no renegotiations took place. In fact, admitting the possibility of renegotiation can

---

[5] A continuous-time counterpart of one-period contracts would be an infinite number of contracts. This would be rather peculiar, and one should restrict bargaining to certain discrete points in time.

[6] More specifically, individual rationality must be satisfied in each spot contract whereas in the long-term contract, individual rationality is required only in the overall contract. The latter is a weaker requirement than the former and long-term contracting hence allows for intertemporal transfers of payoff.

endanger an earlier agreement if it so happens that renegotiation is beneficial to all players[7].

The purpose of the paper is to address, in a dynamic marketing channel context, a number of the above general issues of cooperation. The paper progresses as follows. Section 2 presents the differential game. Section 3 identifies a feedback Nash equilibrium, one candidate for an disagreement equilibrium solution. Section 4 provides the basic ingredients of a cooperative solution.

2   The model

The channel faces no competition and carries no inventories. All consumer sales are made by the retailer. Decision variables are the firms' advertising expenditures and prices (transfer price and retail price, respectively). Let t denote calender time, $t \in [0,T]$. The horizon date T is fixed and finite. Define

$p_M(t)$:   transfer price charged by the manufacturer per unit of sales to the retailer[8].

$p_R(t)$:   unit retail price paid by consumer to the retailer.

$q(t)$:   rate of consumer sales at time t.

$a_j(t)$:   player's $j \in \{M, R\}$ rate of advertising expenditure at time t.

---

[7] The renegotiation problem has attracted interest in repeated game theory. A defector could propose the other players to go back to cooperation instead of having them implementing their punishments, if punishment hurts the punishing players as much as they hurt the defector. If the punishing players cannot resist the temptation to renegotiate, a trigger strategy-supported cooperative outcome is deemed to be unstable. Renegotiation-proof threats deter any defection and are not Pareto dominated by any other threat that deters defection.

[8] Thus, we consider a linear contract. The retailer simply pays the manufacturer an amount proportional to the amount of goods purchased. A nonlinear contract could involve a franchise fee, in which case a constant amount would be added to the purchase cost.

$G(t)$:   retailer's stock of consumer advertising goodwill
                     by time t.

c: manufacturer's constant unit cost of production.

The consumer sales rate depends on the retail price and the stock of consumer goodwill $q = f(p_R, G)$ where $\partial f / \partial p_R < 0$, $\partial f / \partial G > 0$.

We choose a multiplicatively separable relationship, $q = g(p_R) h(G)$ (Fershtman et al. (1990)). For $g(p_R)$ and $h(G)$ we assume linear functions (Horsky (1977), Rao (1984)). Schmalensee (1972) notes that unless sales are linear in G, it is impossible to remove the unobservable G when estimating the demand parameters. Thus, consumer sales are given by

$$q(t) = \gamma G(t) [\alpha - \beta p_R(t)] \tag{1}$$

A key variable of our model is the retailer's stock of consumer goodwill. Chintagunta and Jain (1992) studied a model having some similarities to ours, but assumed that each firm has its own, individual goodwill stock which can only be influenced by the firm's own effort (advertising expenditure). Chintagunta and Jain (1992) employed the Nerlove-Arrow (1962) accumulation dynamics for the two goodwill stocks: $\dot{G}_M = a_M - \delta G_M$, $\dot{G}_R = a_R - \delta G_R$. They assumed that the retail sales rate is an increasing function of both goodwill levels (see also Fershtman et al. (1990)). Hence, consumers increase their demand not only when the retailer stock of goodwill increases, but also when the manufacturer stock increases.

We suppose that consumer goodwill is retailer-specific and it is this single stock of goodwill that affects consumer demand. The manufacturer may contribute to the build-up of consumer goodwill by implementing advertising efforts targeted toward the consumers. For goodwill dynamics we use a straightforward extension of the Nerlove-Arrow model:

$$\dot{G}(t) = a_M(t) + a_R(t) - \delta G(t) \quad ; \quad G(0) = G_o = const. > 0 \,(2)$$

Advertising costs are assumed to be quadratic in the advertising rates: The advertising cost of player $j \in \{M,R\}$ equals $\dfrac{w_j}{2} a_j^2$.

The objective functional of a firm consists of the undiscounted profit stream over the planning period plus a salvage value. We disregard discounting: including positive discount rates adds only little to the paper. Salvage values are quadratic functions of the terminal stock of goodwill. This reflects an assumption that the channel members put increasing emphasis on a large terminal stock of goodwill. The assumption is open to discussion but could be plausible if the channel foresees the entry of a competing channel at, or shortly after, the terminal instant T. A large terminal stock of goodwill would give the channel a competitive advantage, or may be even deter the entry of the competitor. To define the individual objective functionals, let S_M and S_R be positive constants.

$$J_M = \int_0^T [(p_M(t) - c)\gamma G(t)[\alpha - \beta p_R(t)] - \frac{w_M}{2} a_M(t)^2]dt + S_M G(T)^2$$

$$J_R = \int_0^T [(p_R(t) - p_M(t))\gamma G(t)[\alpha - \beta p_R(t)] - \frac{w_R}{2} a_R(t)^2]dt + S_R G(T)^2$$

3   Noncooperative equilibria

We derive price and advertising strategies for the two channel members when each member independently maximizes his decision variables. This situation is referred to as "within-channel competition" or "channel conflict" and occurs when a manufacturer disagrees with his own dealer(s). The resulting noncooperative profits show the effects of lacking communication and cooperation. They also serve as lower bounds for the shares of total channel profits that channel members would accept in a coordinated solution.

Various assumptions about the roles of the channel members, and their use of information, are available. When channel members make their decisions simultaneously and uncoordinated, a Nash equilibrium could be an appropriate notion to predict the outcome of intra-channel rivalry. Two different information structures are usually adopted. A Markovian Nash equilibrium (feedback Nash equilibrium) is an equilibrium in which channel members employ strategies that depend on current time t and current state $G(t)$. A crucial assumption here is that both members are supposed to know the retailer's stock of consumer goodwill at any instant of time. On the other hand, employing open-loop strategies means that both players precommit, at the start of the game, to apply fixed time-functions for their decisions. In sequential play, a classical asymmetric solution concept is a Stackelberg equilibrium in which one of the firms takes the role as the leader who is the first one to announce his decisions. The follower reacts rationally upon the leader's choices. Feedback or open-loop information can also be employed in a Stackelberg game. We consider here a simultaneous play setup with feedback information structure.

Since prices are absent from the dynamics, a static pricing game at each $(t, G(t))$ is played:

$$\text{Manufacturer: } \max\{\pi_M = (p_M(t)-c)\gamma G(t)(\alpha-\beta p_R(t))/$$
$$c \leq p_M(t) \leq \bar{p}_M\}$$

$$\text{Retailer: } \max\{\pi_R = (p_R(t)-p_M(t))\gamma G(t)(\alpha-\beta p_R(t))/$$
$$p_M(t) \leq p_R(t) \leq \frac{\alpha}{\beta}\}$$

(3)

The maximands are individual instantaneous profits, net of advertising expenditure. Since the manufacturer's profit is linear in the transfer price, we impose an upper bound, $\bar{p}_M$, on transfer price. The optimal transfer price is constant, equal to $p_M = \bar{p}_M$ whenever $q > 0$. (For $q \leq 0$ the transfer price is indeterminate but also irrelevant). The boundary solutions for the retail price seem unlikely, yielding the retailer zero instantaneous profits, and we confine our interest to interior solutions. An optimal retail price is unique and satisfies

$$p_R^N = \frac{1}{2\beta}[\alpha + \beta \bar{p}_M] \tag{4}$$

where the superscript "N" identifies the Nash game. Thus, optimal retail price is constant but depends on the maximal transfer price[9]. It is reasonable to assume $\alpha > \beta \bar{p}_M$, implying $\alpha > \beta c$. Introduce the constants

$$\theta_M^N = (\bar{p}_M - c)\gamma(\alpha - \beta p_R^N) > 0$$

$$\theta_R^N = (p_R^N - \bar{p}_M)\gamma(\alpha - \beta p_R^N) > 0 \tag{5}$$

The firms determine their dynamic advertising rates as Markovian decision rules, i.e., $a_j^N = \phi_j(t, G(t))$, $j \in \{M, R\}$. We state the basic conditions for strategy pair $\phi_M(t, G(t))$, $\phi_R(t, G(t))$ to yield a Markovian Nash equilbrium, leaving the details for the reader. There exist continuously differentiable functions $V^j : \Re_+ \times [0, T] \rightarrow \Re$ such that the following Hamilton-Jacobi-Bellman (HBJ) equations are satisfied for all $(t, G) \in [0, T] \times \Re_+$

---

[9] This is a simple implication of the stationarity of the pricing game. Letting the parameters (e.g., the intercept) of the demand function depend on time would produce a time-varying retail price.

$$\max \ \{\theta_M^N G - \frac{w_M}{2} a_M^2 + \frac{\partial V^M}{\partial G} [a_M + \phi_R(G(t),t) - \delta G] /$$

$$a_M \geq 0\} = - \frac{\partial V^M}{\partial t}$$

(6)

$$\max \ \{\theta_R^N G - \frac{w_R}{2} a_R^2 + \frac{\partial V^R}{\partial G} [\phi_M(G(t),t) + a_R - \delta G] /$$

$$a_R \geq 0\} = - \frac{\partial V^R}{\partial t}$$

The boundary conditions

$$V^j(G,T) = S_j G^2 \ ; \ j \in \{M,R\} \tag{7}$$

must hold for all $G \geq 0$, and $a_j = \phi_j(G(t),t)$ maximizes the left-hand side of the HBJ-equations. Performing these maximizations yields, due to concavity of the maximands with respect to a_j, a unique pair of advertising rates

$$a_j^N = \frac{1}{w_j} \frac{\partial V^j}{\partial G} \ , \ j \in \{M,R\} \tag{8}$$

Equation (8) has an economically intuitive interpretation: set advertising at the rate at which marginal advertising cost, $w_j a_j^N$, equals the marginal increase in a firm's optimal profits, coming from a marginal increase of the stock of goodwill. Note $a_j^N = 0$ $\rightarrow$ $-w_j a_j^N + \partial V^j/\partial G < 0$ which is intuitive. Inserting from (8) on the left-hand side of (6) yields

$$\theta_j^N G + \frac{1}{2 w_j} (\frac{\partial V^j}{\partial G})^2 - \delta G \frac{\partial V^j}{\partial G}$$

$$+ \frac{1}{w_i} \frac{\partial V^j}{\partial G} \frac{\partial V^i}{\partial G} = - \frac{\partial V^j}{\partial t} ; \ i,j \in \{M,R,\}; \ i \neq j$$

(9)

Suppose that value functions are given by

$$V^j(G,t) = \kappa_j(t) + \sigma_j(t)G + \mu_j(t)G^2 \qquad (10)$$

in which the time-functions κ_j, σ_j, and μ_j are continuously differentiable. The boundary conditions (7) then require

$$\kappa_j(T) = \sigma_j(T) = 0 \ , \ \mu_j(T) = S_j \qquad (11)$$

Partial differentiation of V^j in (10) yields

$$\frac{\partial V^j}{\partial G} = \sigma_j(t) + 2\mu_j(t)G \quad , \quad \frac{\partial V^j}{\partial t} = \dot\kappa_j(t) + \dot\sigma_j(t)G + \dot\mu_j(t)G^2 \qquad (12)$$

which shows that the Markovian advertising strategies are linear decision rules in terms of the goodwill stock:

$$\phi_j(G,t) = \frac{1}{w_j}[\sigma_j(t) + 2\mu_j(t)G] \qquad (13)$$

Depending on sign (μ_j), advertising may increase or decrease with goodwill stock at time t. Substitute from (16) and (17) into (13) to see that if we can find time-functions κ_j, σ_j, and μ_j that satisfy a system of six ordinary differential equations, then the hypothesized value functions (10) do satisfy the Hamilton-Jacobi-Bellman equations. This is indeed possible: the derivations are omitted.

4 A cooperative solution

The firms wish to coordinate their decisions in order to maximize a weighted sum of their respective profits over the duration of the game. The joint profit functional is given by

$$\pi = \int\limits_0^T [\rho_M\{(p_M(t)-c)\gamma G(\alpha-\beta p_R(t)) - \frac{w_M}{2}(a_M(t))^2\} +$$

$$\rho_R\{(p_R(t)-p_M(t))\gamma G(\alpha-\beta p_R(t)) - \frac{w_R}{2}(a_R(t))^2\}]dt + \tag{14}$$

$$[S_M+S_R]G(T)^2$$

$$\rho_M, \rho_R > 0, \; \rho_M + \rho_R = 1. \tag{15}$$

We have an optimal control problem consisting of (14) and the state dynamics. It is known that a quadruple of prices and advertising rates is Pareto efficient if there exist feasible bargaining weights ρ_M and ρ_R, and the quadruple maximizes the objective (14).

Consider the (static) problem of determining prices at each instant of time. Differentiation of the Hamiltonian yields

$$\frac{\partial H}{\partial p_M} = [\rho_M - \rho_R]\gamma G(\alpha-\beta p_R) = [\rho_M - \rho_R]q \tag{16}$$

If q is zero, transfer price is irrelevant. If q is positive, the relationship between the bargaining weights completely determines the transfer price:

$$\rho_M \begin{bmatrix} > \\ < \end{bmatrix} \rho_R \rightarrow p_M^C(t) = \begin{bmatrix} \bar{p}_M \\ c \end{bmatrix} \tag{17}$$

in which the superscript "C" denotes a cooperative solution. The two-value transfer price policy is a product of a linearity of the model. (For a similar result, see Benhabib and Ferri (1987)). If the manufacturer has the larger bargaining weight, he quotes his maximal transfer price, and vice versa. If weights are equal, the transfer price is indeterminate. To avoid the extreme outcomes $\bar{p}_M$ and c, which seem less likely in a cooperative mode of play, we assume equal weights. This turns the problem into the "joint profit maximization" problem but leaves the transfer price indeterminate (since it cancels out in the joint objective). The cooperative retail price is constant and uniquely given by

$$p_R^C = \frac{1}{2\beta}\{\alpha + \beta c\} \tag{18}$$

As expected, this price is lower than the retail price in the noncooperative case (cf. (4)). Comparing the outcomes of noncooperative and cooperative pricing shows that at any $(t, G(t))$ the firms can increase the aggregate instantaneous profit (net of advertising expenditures) if they set their prices cooperatively. Transfer price is indeterminate but can be determined to allocate the surplus in any particular way. Dividing the surplus such that each firm receives strictly more than its noncooperative profit is a Pareto improvement and shows (not unexpectedly) that the noncooperative Nash outcome is inefficient. Denote the costate variable by $\lambda(t)$. Optimal cooperative advertising rates are as follows:

$$a_j^C(t) = \max\{0, \frac{\lambda(t)}{w_j}\} \tag{19}$$

When advertising rates are positive we have

$$\frac{a_M^C}{a_R^C} \begin{bmatrix} > \\ < \end{bmatrix} 1 \;\; for \;\; w_R \begin{bmatrix} > \\ < \end{bmatrix} w_M$$

which simply says that the firm having the advertising cost advantage will do the majority of the advertising. Define

$$\theta_M^C = (p_M - c)\gamma(\alpha - \beta p_R^C), \;\; \theta_R^C = (p_R^C - p_M)\gamma(\alpha - \beta p_R^C). \tag{21}$$

The costate equation becomes

$$\dot{\lambda} = \delta\lambda - [\theta_M^C + \theta_R^C] \;; \;\; \lambda(T) = 2G(T)[S_M + S_R] \tag{22}$$

Using (22) it is easy to establish that the costate is strictly positive for all $t \in [0,T)$. The optimal advertising rates are strictly positive for $t \in [0,T]$ and decrease over time.

A coordinated policy is a quadruple $(p_M^C(t), p_R^C(t), a_M^C(t), a_R^C(t))$ in which p_M^C is indeterminate, p_R^C is constant and given by (20), and $a_j^C(t)$ is given by (21). A coordinated policy is agreeable at $(0, G_0)$ if, at any feasible $(t, G(t))$ along the agreed path, the prolongation of the coordinated policy for each firm payoff dominates the prolongation of the noncooperative solution $(\bar{p}_M, p_R^N(t), a_M^N(t), a_R^N(t))$. The noncooperative retail price p_R^N is constant and given by (4); noncooperative advertising rates $a_j^N(t)$ are given by (8) and (13). Denote by $\Sigma_j(G(t),t)$ the share of total channel profits received by firm $j \in \{M, R\}$ when both firms apply their coordinated strategies over the time interval $[t,T]$. At any point along the agreed trajectory the total cooperative profit must be no less than the sum of the noncooperative profits. Moreover, the individual dominance relations must hold with strict inequality for at least one j

$$\Sigma_j(G(t),t) \geq \int_t^T [\theta_j^N G(s) - \frac{w_j}{2}(a_j^N(s))^2]ds + S_j G(T)^2; \quad (23)$$

$$j \in \{M,R\} \; ; \; \forall t \in [0,T]$$

for all feasible G along an agreed cooperative path. The shares Σ_j must satisfy for all $t \in [0,T]$

$$\Sigma_M(G^C\!(t),t) + \Sigma_R(G^C\!(t),t) = \int_t^T [(p_R^C - c)\gamma G^C\!(s)(\alpha - \beta p_R^C)$$

$$- \frac{1}{2}[w_M(a_M^C\!(s))^2 + w_R(a_R^C\!(s))^2]ds + (S_M + S_R)G(T)^2 =$$

$$\int_t^T [\pi^C\!(s) - \frac{w_M + w_R}{2w_M w_R}\lambda(s)^2]ds + (S_M + S_R)G(T)^2$$

in which $\pi^C\!(s) = \dfrac{\gamma G^C\!(s)}{4\beta}(\alpha - \beta c)^2$, and λ is the unique solution of (22). Obviously, the outcome of the negotiations is still indeterminate: any division (Σ_M, Σ_R) of the joint profit that satisfies the imposed conditions is a possible outcome.

5   Concluding remarks

This paper has discussed some basic issues that should be addressed when a manufacturer and a retailer wish to coordinate their actions to avoid channel inefficiencies. We have exploited some, but far from all, of these notions in the context of a differential game of pricing and advertising. Some results concerning a feedback Nash equilbrium, an open-loop Nash equilibrium, and a coordinated solution were reported. Future work will concern in particular the Stackelberg games and the problem of dividing the joint profits in a coordinated solution.

References

Benhabib, J. and G. Ferri (1987), "Bargaining and the Evolution of Cooperation in a Dynamic Game", *Economics Letters*, 24, 107-111.

Chintagunta, P.K. and D. Jain (1992), "A Dynamic Model of Channel Member Strategies for Marketing Expenditures", *Marketing Science* 11, 2, 168-188.

Corfman, K.P. and S. Gupta (1993), "Mathematical Models of Group Choice and Negotiation", in J. Eliashberg and G.L. Lilien (Eds.), *Handbooks in Operations Research and Management Science: Marketing*. North-Holland, Amsterdam.

Coughlan, A.T. (1985), "Competition and Cooperation in Marketing Channel Choice: Theory and Application", *Marketing Science*, 4, 2, 110-129.

Coughlan, A.T. and B. Wernerfelt (1989), "On Credible Delegation by Oligopolists: A Discussion of Distribution Channel Management", *Management Science*, 35, 2, 226-239.

Fertshtman, C., V. Mahajan, and E. Muller (1990), "Market Share Pioneering Advantage: A Theoretical Approach", *Management Science*, 36, 8, 900-918.

Gerstner, E. and J.D. Hess (1995), "Pull Promotions and Channel Coordination", *Marketing Science*, 14, 1, 43-60.

Horsky, D. (1977), "An Empirical Analysis of Optimal Advertising Policy", *Management Science*, 23, 10, 1037-1049.

Jeuland, A.P. and S.M. Shugan (1983a), "Managing Channel Profits", *Marketing Science*, 2, 3, 239-272.

Jeuland, A.P. and S.M. Shugan (1983b), "Coordination in Marketing Channels", in D. Gautschi (Ed.), *Productivity and Efficiency in Distribution Systems*, Elsevier, New York, 17-38.

Jeuland, A.P. and S.M. Shugan (1988), "Channel of Distribution Profits When Channel Members Form Conjectures", *Marketing Science*, 7, 2, 202-210.

Johansson, J. and L.-G. Mattsson (1988), *Internationalization in Industrial Systems: A Network Approach*, Croom-Helm, London.

Kohli, R. and H. Park (1989), "A Cooperative Game Theory Model of Quantity Discounts", *Management Science* 35, 6, 693-707.

McGuire, T.W. and R. Staelin (1986), "Channel Efficiency, Incentive Compatibility, Transfer Pricing, and Market Structure", in *Research in Marketing*, Vol. 8, JAI Press Inc., 181-223.

Moorthy, K.S. (1987), "Managing Channel Profits: Comment", *Marketing Science*, 6, 4, 375-379.

Moorthy, K.S. (1988), "Strategic Decentralization in Channels", *Marketing Science*, 7, 4, 335-355.

Nerlove, M. and K.J. Arrow (1962), "Optimal Advertising Policy under Dynamic Conditions", *Economica*, 39, 129-142.

Rao, R.C. (1984), "Advertising Decisions in an Oligopoly: An Industry Equilibrium Analysis", *Optimal Control Applications and Methods*, 5, 4, 331-344.

Rey, P. and B. Salanie (1990), "Long-Term, Short-Term and Renegotiation: On the Value of Commitment in Bargaining", *Econometrica*, 58, 3, 597-619.

Schmalensee, R. (1972), *The Economics of Advertising*, North-Holland, Amsterdam.

Shaffer, G. (1995), "Rendering Alternative Offerings Less Profitable with Resale Price Maintenance", *Journal of Economics & Management Strategy*, 3, 4, 639-662.

Shugan, S.M. (1985), "Implicit Understandings in Channels of Distribution", *Management Science*, 31, 4, 435-460.

Shugan, S.M. and A.P. Jeuland (1988), "Competitive Pricing Behavior in Distribution Systems", in T.M. Devinney (Ed.), *Issues in Pricing*, Lexington Books, 219-237.

Stern, L.W. and A.I. El Ansary (1977), *Marketing Channels*. Prentice-Hall, Englewood Cliffs, NJ.

Zusman, P. and M. Etgar (1981), "Marketing Channel as an Equilibrium Set of Contracts", *Management Science*, 27, 3, 284-302.

Avoiding Myopia: Seeing The Competition

George E. Cressman, Jr.
Marketing Development Group, DuPont, Wilmington, Delaware, U.S.A.

Abstract: Levitt(1975) argues that some firms lose market position because they define their business too narrowly. Levitt calls this narrow vision "Market Myopia." More recently, Day and Nedungadi(1994) suggest that managers selectively focus only on parts of market phenomena, and thus focus both on the search for and interpretation of market information.

While these observations may explain the observed behavior of managers, they leave open the question of why managers behave this way. This paper argues an underlying causal factor of both narrow business definition and selective competitor views are restricted competitor boundaries. This paper suggests scoping competitor boundaries through four stages:
- Industry to industry competition for the attention, loyalty and spending of consumers.
- Within the industry, between industry levels, competition for a share of the revenue flowing into the industry.
- Strategic group to strategic group competition at a specific level within an industry.
- Within strategic groups, firm to firm competition among participants of a specific group.

This paper proposes that myopic firms see competition only as the last level, i.e., as occurring only at the within strategic groups, or firm to firm, level. A more holistic view sees competition at all levels and focuses at all four levels.

This more holistic view significantly impacts managerial thinking. At its broadest level, management must focus on winning through their industry. This implies that firms must focus on industry success in providing time, place, and functional utility for the end consumer marketplace.

The work developed in this paper proposes a framework for developing a four stage competitive view. Suggestions for managerial action to implement this view are offered.

MARKETING MYOPIA: SEEING THE COMPETITION

Introduction

Theorists have argued that the ultimate criterion of organizational performance is long term survival and growth(see Chandler 1962; Katz and Kahn 1966; and Thompson 1967). To achieve this, organizations align with their environments to remain competitive and innovative(Barnard 1938; Lawrence and Lorsch 1967; and Thompson 1967). Thus, a key premise of competitive management strategy development compels an attempt to align the organization and its operating environment over time. The goal of this alignment attempt is to assure competitiveness and the longer term survival of the firm(Hambrick 1983; Summers 1980). Indeed, Ansoff(1965) has argued that the role of strategy, and presumably competitive strategy, is to reconcile the organization with its environment.

Alignment implies that the firm must engage in actions which create the potential to learn, unlearn and/or relearn successful competitive strategies from past behaviors(Fiol and Lyles 1985). Here, learning enables organizations to build on an the organization's collective understanding and interpretation of their environment and to begin to assess viable competitive strategies to cope with this environment(Daft and Weick 1984; and Donaldson and Lorsch 1983).

In a widely quoted paper, Levitt(1975) asserts that businesses stop growing and begin to decline because their managers define their business not in terms of customer needs served, but in terms of products or services offered. Thus, managers fail to see ways to continuously alter their offerings to serve more of customers' needs, and are likewise inadequately aware of potential substitute offerings until these substitutes displace them. This failure to correctly "define" the business became the basis of Levitt's "Marketing Myopia." In a more general sense, this might be seen as a narrow definition of the appropriate operating arena.

While Levitt's assertion that businesses mature and decline because managers fail to properly define their operating arena is a useful insight, Levitt offers little guidance on how managers are to correct the narrow definition fault. It is the central premise of this paper that another view of the cause of *Marketing Myopia* is a failure of managers to correctly define their competitive arena. It is argued here that managers must broaden their view of the competitive arena in order to improve both their view of served customer needs and the generation of longer term viable strategies. This paper suggests that it is in the broader view of competition suggested here that an improved organization-environment fit is achieved.

Day and Nedungadi(1994) build on the literature stream(see Barnes 1984; Daft and Weick 1984; Kiesler and Sproull 1982; Smircich and Stubbart 1985; and Stubbart 1989) centered on managerial use of mental models to describe their operating environment to suggest that competition is the result of managerial representations of their world. This suggests that competitive arenas are created by managerial assessments of competitive encounters; competitive arenas are less the product of some natural set, but are rather the creation of a managerial process to interpret and structure what is perceived in the world. This paper builds on this concept by suggesting a broader view of the competitive arena. This view is rooted in ultimate consumer markets and the choices consumers make for the uses of their disposable income. Thus a more traditional view of firm to firm competition is enhanced by developing a view of consumer use driven competition.

Competitor Identification

Markets

The competitive assessment starts with end use consumers in a market(see Figure 1). Here a market is defined as:

Market - A collection of end use consumers who are willing to spend money on the satisfaction of wants and needs. Every market has multiple choices available among which consumers make decisions about the use of their disposable income. For example, consumers may view continuing education in a university setting to trade off entertainment options as they compete for a consumer's desire to enhance self image.

The presence of multiple choices is a crucial part of characterizing markets. While we may be tempted to think of "monopolistic" suppliers to markets, these exist only in conceptual terms. For example, electrical power generating firms are often viewed as monopolies, but in fact, these firms face the following competition:

- Functional substitutes - oil or gas.
- Not-in-kind substitutes - extra insulation, more clothing, blankets(all of which may reduce the demand for electrical power).
- Alternatives - conservation(which also reduces demand).

A similar set of alternatives exists whether the electrical power is used for:

- In home entertainment - competition might be a recreational activity.

179

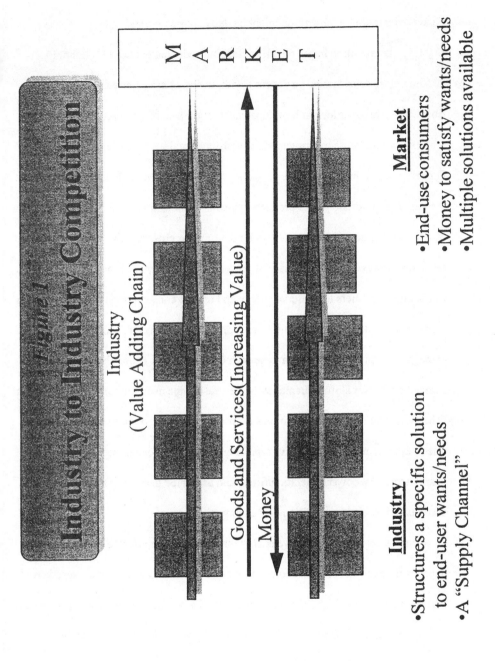

Figure 1

Industry to Industry Competition

Industry
(Value Adding Chain)

Goods and Services (Increasing Value)

Money

M
A
R
K
E
T

Market
- End-use consumers
- Money to satisfy wants/needs
- Multiple solutions available

Industry
- Structures a specific solution to end-user wants/needs
- A "Supply Channel"

- Food refrigeration - competition might be daily shopping trips.

- Lighting - competition might be oil, or sleeping(reducing the need for lighting).

While some of these examples may seem trivial, the practicing manager should:

- Encourage a broader look at how consumers might use their discretionary income, thus exploring a more extensive look at consumer options for the use of their income. For example, in the early 1970s, generation of electricity through nuclear power was seen as an appropriate power production means in the United States. This was driven by a belief that demand for power would continue to grow rapidly. Two successive oil crises, however, drove the price of power generation by all means higher, increasing consumers to conserve energy and reduce consumption. This reduction in demand was aided by tax incentives for individuals to reduce consumption. Nuclear plant generation of power began to be seen as too expensive, and the growth of demand for power slowed dramatically, decreasing the need for nuclear power plants. (The perceived environmental threats from nuclear plants, of course, also played a role in decreasing demand for this type of energy). These events seem to have been unanticipated by power generating firms, and as argued here, this "myopia" was accentuated by a poor definition of the competitive arena.

A further act in the alternative energy play involves the effort to convert oil shale deposits to readily expendable forms of energy. These projects were premised on high, and continuing high, prices for crude oil. These high prices were not sustained, in large part due to the actions of non-OPEC oil producing countries. Again, a

narrow view of the competitive arena seems to have contributed to a failure to
anticipate these occurrences.

- Challenge organizationally accepted mental models of consumer spending choices.
  A broader view of the competitive set can enhance the firm's understanding of how
  consumers make purchase decisions, and how these decisions impact managerial
  actions.

The rationale for this broader look at consumer options is that consumers chose
solutions for their wants/needs on the basis of the value delivered(benefits received) for
the cost of acquiring the benefits. If a particular solution available in a market provides
an inadequate level of benefits, or provides adequate benefits at competitively
disadvantageous costs, relative to other recognized solutions in the market, the
organization's solutions will be ignored in preference to solutions perceived to be more
desirable. Competitors must pay attention to the perceived value delivery vs. acquisition
costs to ultimate consumers, or risk consumer migration to perceived more advantageous
offerings. At a minimum, competitors should evaluate the value delivery of all actual
and potential solutions and position themselves to participate in those consumers
perceive to be the more advantageous.

A good example of spending choices is the decisions business customers make
regarding air travel. A common purpose for business travel is to communicate with other
business people or customers on a face to face basis. Nevertheless, air travel is an
expensive process when considered in the context of the opportunity cost of the business
person's time. Given the challenging competitive arenas many firms face, and the

smaller executive ranks resulting from downsizing, time is an exceedingly precious commodity. Thus, any process which might allow face to face communication while reducing time consumption is likely to win market share.

Video conferencing provides close substitutes for these needs. Thus, increasingly, commercial airline operations compete with global telecommunications firms. To the extent airline carriers see part of their competitive set as firms facilitating communication, they may be able to better understand competition and better construct need satisfying solutions.

Industries

The multiple solutions consumers see in the market are created by Industries:

Industry - An industry is a value adding chain which creates a specific solution to end-user wants/needs.

As goods and services move closer to ultimate consumers in value adding chains, value is created in the form of delivered, and hopefully desired, benefits. These delivered benefits have the following components:

- Tangible - physical products, e.g., an automobile.

- Intangible - services, e.g., warranty or maintenance of the automobile.

- Emotive - image and positioning , e.g., the automobile advertising which creates some image about the automobile.

Within the Industry, there are three kinds of flows:

- Value added increases as the solution moves towards the consumer, increasing tangibles are added to the offering, services are used to augment the tangible offering, and as image is created around the solution. This value could be negatively perceived by the ultimate consumer if the consumer does not want the tangible elements, service elements do not support customer needs, and/or the image is confusing to consumers.

- Money flows from ultimate consumers towards more fundamental raw material suppliers within the value adding chain.

- Information flows both from ultimate consumers towards more fundamental raw material suppliers and in the reverse direction. Information is both supplied and consumed in the transactions occurring in the industry.

Types of Competition

The market-industry construct allows drawing a view of competition at four levels:

- In the market *between* industries. This is the battle for consumers' attention and disposable income. This battle is decided on differences in delivered benefits vs. the cost of acquiring these benefits.

- *Within* the industry. This competition is for shares of the revenue flow within the industry. Note that if the industry's solution is not well regarded by consumers, the industry will have an increasingly smaller total sum of money to divide(that is, compete for) within the industry. This battle is waged both on the contribution of individual levels to ultimate consumer needs and on the basis of power held by individual levels in the industry.

- *Between* strategic groups within a specific level of the industry. A strategic group is defined as a set of competitors which have similar or very closely matched strategic approaches to the industry and market. For example, among commercial airline operators in the United States, one strategic group(consisting of American, Delta and United Airlines) focuses primarily on business travelers while another focuses primarily on the pleasure traveler. At a specific level within an industry, each strategic group competes for revenues flowing to that industry level at the expense of other strategic groups.

- *Within* a strategic group. This is the most direct form of competition. Here, each individual tries to maximize the revenues flowing to itself. Each firm intends for others within its strategic group to receive less of the revenues flowing to the strategic group. This is the more traditional firm to firm competitive level.

Assessing The Competition

The first step in competitor assessment, then, is to determine what choices consumers have(potential choices) and perceive they have(perceived choices) for use of their discretionary income, and which trade-offs consumers are making among these choices. The distinction between potential choices and perceived choices is important because:

- Perceived choices comprise the choice set consumers are using currently for choosing among options. This may not be a universal set of choices for numerous reasons(e.g., inadequate market communication(consumers unaware of possible choices) or ignored communications). In some markets, a frequent reason for an option not residing in the perceived choice set is the option is perceived as too expensive(either in monetary or non-monetary terms).

- Potential choices are the choices consumers might make in the future. In order for potential choices to enter the perceived choice set, consumers must be aware of the option, be able to acquire the choice, and be motivated to acquire the choice.

This first step assessment takes the following form:

* Determine what market(s) are of interest.

* For each market, determine consumer wants and needs, and the total funds spent to satisfy these wants and needs.

* Define alternative solutions for wants/needs satisfaction, and the funds spent on each solution. In this analysis, particular attention should be focused on emerging or potential solutions; these may develop into future competition.

* For each solution(current or emerging/potential), assess:

      - Desirability to consumers,

      - Delivered benefits, as perceived by consumers, and

      - Perceived acquisition costs.

The most significant competitors at this level are those solutions(that is, industries) which have good solutions when viewed from a consumer perspective. At this level, consumer perceived value is assessed as,

Value To	=	Delivered	-	Costs to Acquire
Consumers		Benefits		Delivered Benefits

This same concept could be applied to both existing and emerging/potential solutions(industries). Emerging/potential solutions may appear to have high acquisition costs as they are introduced; in this case the rate of acquisition cost decline(e.g., declining as experience effects accumulate) should also be explored to determine if/when these newer solutions are likely to be perceived by consumers as viable offerings.

An important component of analysis at this level is a dynamic analysis. Relative rates of change among solutions in delivered benefits and/or acquisition costs can dramatically alter the positions of competing offerings as perceived by consumers. For example, when older, established technologies compete with newer technologies, perceived cost differences may be altered by rapidly declining costs for the new technologies. Here it is the relative change that is important.

Assessment of competition among solutions for a share of consumer disposable income results in a flow of this income into some or all of the industries providing competing solutions. How much disposable income flows to specific industries is a function of the consumer perceived viability of the competing solutions. This flow of disposable income to the industries becomes revenue available to successfully competing industries.

A product of defining the competitive set at the consumer market level is an understanding of the criteria consumers use to make choices among competing solutions. These criteria become the critical wants/needs consumers are seeking to satisfy within the market. Contribution to the satisfaction of these critical wants/needs becomes the basis for competition within the industry.

Industry Dynamics

Having developed a view of competition at the consumer level, the second stage of competition becomes competition for the revenue flowing through the industry. In order to understand competition at this level, an industry diagram consisting of the value adding steps within the industry must be generated. This value adding chain flows from fundamental raw materials to the ultimate goods/services consumed within the market.

Individual levels within the industry compete in contributing to generating solutions for consumer critical wants/needs. Those firms which make greater contributions to the generation of solutions which visibly provide critical wants/needs are stronger competitors for the revenue flowing into the industry. Thus, because they structure the final tangible offering seen by ultimate consumers, automobile manufacturers typically command large portions of the value flowing through their industry.

Individual industry levels engage in activities which provide contributions to generating solutions for consumer critical wants/needs. These activities fall into four general categories(see Table 1): Transformation, Logistics, Financial, and Marketing activities. The degree to which these activities are effectively attached to the consumer critical wants/needs determines the desirability of the outcomes of these activities to consumers. Industry levels which are more effective in attaching their activities to consumer critical wants/needs will attract more of the revenues flowing through the industry. Firms which are more cost efficient in executing these activities will retain more of the revenue flowing through their industry level.

Table 1

Industry Activities

Transformation

Manufacturing
Converting
Fabricating
Assembling
OEM or system assembly
Repackage
Break bulk
Regrade products
Accumulate/aggregate assortments
Resize lots
Provide warranty

Financial

Buy(own) product
Receive orders
Issue invoices
Collect payments
Issue credits
Assume acc. receivable risk
Assume prod. obsolescence risk
Assume prod. deterioration risk
Consign stock

Distribution

Provide special transportation
Provide special warehousing
Store product
Ship product
Expedite shipments
Maintain inventories

Marketing

Develop new business
User service/support
User training
Evaluate/understand user needs
Provide product information
Gather feedback or information
Advertising/promotion

Thus, competition within the industry takes two primary forms:

- Competition centered in effectiveness, that is, competition centered around those activities which make the most contribution to consumer critical wants/needs. Within the industry, more effective industry levels attract greater shares of the revenues flowing within the industry.

- Competition centered in efficiency, that is, competition centered around retaining shares of revenue flowing to a given level through aggressive cost control and the use of power to maintain position.

A useful mechanism for evaluating the level of competition within an industry is to explore reward/effort indices for the individual levels of the industry. Reward is defined as the revenue retained by the individual level of the industry, while effort is defined as the costs of executing the activities within the industry level. This index(the R/E index) for each industry level can have three values:

* R/E < 1 This level's costs exceed its retained revenue flow. This could be the result of excessive non-value adding activities, poor cost control, or few contributions to consumer critical wants/needs. Regardless of how this occurs, these levels may become powerful competitors for industry levels further upstream of them because they attempt to force price competition among their suppliers.

* R/E = 1 This level is operating at break-even. In order to improve its position, it must increase its contribution to consumer critical wants/needs(attracting more industry revenue flow from downstream industry competitive levels) or reduce its costs(attracting more industry flow from upstream industry competitive levels).

* R/E > 1 This level has achieved a powerful position because it is able to extract revenue flows beyond the cost of its activities. Other industry levels will compete to attempt to attract some of the revenues flowing to this level.

This portion of the competitive assessment provides insights into the following:

♦ The industry's overall contribution to consumer critical wants/needs, and which industry levels are generating this contribution.

♦ Industry level profitability, as measured in the R/E index. Of particular concern with the R/E index are:

  - Industry levels which are not efficient, and have R/E figures less than 1. These levels are likely to force upstream levels to price compete(i.e., to compete for lower shares of the revenues flowing into the industry). This price competition may ultimately force attempts to reduce costs which, when implemented, result in lowered industry effectiveness in contributing to consumer critical wants/needs. This may, in turn, decrease the desirability of the industry's solutions to ultimate consumers.

  - Industry levels with R/E figures greater than 1. These levels extract higher shares of revenues flowing through the industry than do other levels. These lower revenue share levels may compete by trying to cost reduce their activities. Again, this could result in reduction of industry effectiveness.

  - Industry levels with R/E figures close to 1. These figures may indicate levels may be trending towards inefficiency.

Of equal concern in assessing competition at the industry level are changes in the R/E indices. Firms which are aggressively increasing their R/E index are successfully competing for greater revenue shares. This can create very dynamic competitive arenas and signal increasing competitive intensity.

This level of industry competitive analysis gives insights into how revenues reaching the industry will be distributed between levels of the industry. The next stage of the competitive analysis - competition between strategic groups - begins the analysis of how revenues reaching a specific industry level will be distributed.

Strategic Groups

Porter(1980) defines strategic groups as meeting the following criteria:

- Firms within a strategic group have fairly homogenous strategies. The strategic choices these firms make are different from the choices made by firms in other strategic groups.

- The strategic choices made within strategic groups are typically difficult for firms outside the strategic group to duplicate, because of:

  - substantial costs,

  - significant time to duplicate, and/or,

  - uncertainty about the outcome of trying to duplicate others' strategic choices.

As intergroup rivalry emerges between strategic groups, the benefit of properly identifying strategic group membership is in the avoidance of confusing strategic and tactical actions in responding to the initiatives of firms lying in other strategic groups. From a managerial standpoint, the objective of the business is to maximize the flow of

192

revenues to the strategic group of which it is a member, and then to compete with other group members for a greater share of the revenues flowing to the group.

Identification of strategic groups appears more art than science. A typical approach to group identification is to build a series of maps and position competitors on those maps. Maps which show groups of competitors may be further analyzed for the underpinnings of strategic choice that drive the groupings of competitors. Maps which show no grouping patterns are discarded.

In developing maps, the first decision is which dimensioning criteria to use. Porter(1980) suggests the following classes of dimensions for mapping variables:

- Product variety: Physical size

                        Features

                        Inputs employed

                        Performance

                        Product vs. ancillary services

                        Price levels

                        Technology or design

                        Packaging

                        New vs. aftermarket replacement

                        Bundled vs. unbundled

- Buyer type: Business Buyers   Consumers

                        Strategy   Demographics

                        Technological sophistication   Psychographics

OEM vs. user   Lifestyles

Level of vertical integration   Purchase occasion

Size

Ownership

Financial strength

Order pattern

- Channel choice: Direct vs. distribution

Direct mail vs. retail

Distributor vs. broker

Types of distributors/retailers

Exclusive vs. non-exclusive outlets

- Geographic buyer location.

As groups of competitors are identified, underlying patterns of strategic choices must be analyzed. The following questions are useful:

- What common characteristics are apparent among close competitors?

- What discretionary activities do close competitors appear to have in common?

- What customer targets have close competitors set?

- What structural similarities do close competitors have?

- What procedural similarities do close competitors have?

- What technological(product and/or process) similarities do close competitors have?

The next level of analysis considers firm to firm competition within a strategic group.

Specific Competition

The most direct form of competition occurs within a strategic group; on a firm to firm basis. This competition is the level of competition described in more traditional competitor analyses. This competition is for a share of the revenue reaching the strategic group at a specific industry level.

Direct competitor analysis takes the following form:

- What is the competitor's strategic thrust, i.e., what is the competitor trying to do? This assessment should consider competitor thrusts in terms of business management, marketing, manufacturing, finance, logistics, and research/technology.

- What capabilities(enabling forces) are in place that will facilitate the competitor achieving its strategic thrusts? What barriers(restraining forces) will inhibit the competitor's pursuit of its strategic thrusts?

- Given the competitor's strategic thrusts and considering competitor capabilities/ barriers, what is the likely outcome? Will the competitor succeed? Fail? Why?

- Is this likely outcome a threat or opportunity for other competitors? In what way?

In order to succeed, each competitor must have appropriate(to the strategic thrust) capabilities, and no deadly barriers. Considering its strategic thrusts, each competitor must establish capabilities to execute its thrusts. Barriers are the mirror of capabilities; lack of enabling resources, lack of business resolve, poor match of offering to customer demands, poor cost position, etc., are all barriers to accomplishment of a strategic thrust. The presence of these items is a capability that increases the probability a competitor will succeed. The absence of these items is a barrier likely to contribute to the failure of the competitor.

Analysis of firm to firm competition within a strategic group provides insights into the day to day tactical activities of competitive strategy. Often the activities of firm to firm competition are of a more immediate nature, while competition between strategic groups, at the industry level and in the market for a share of consumer disposable income requires a more strategic, longer term perspective.

Conclusions

Successful competitive strategy requires an integrative effort. When seen in its broader perspective, competition consists of strategy postulated simultaneously at four different levels:

- In the market, for a share of consumer disposable income, while competing with other solutions consumer perceive as viable options for the satisfaction of their critical wants and needs.
- Within a value adding chain(an industry) at each level of the industry for a share of the revenues flowing into the industry.
- At specific industry levels, in the form of competition between strategic groups for a share of the revenue flowing into the specific industry level.
- Within a strategic group, at the firm to firm level for a share of the revenue flowing to a strategic group.

The traditional approach to competitive strategy has focused on this last competitive battle, firm to firm competition. Such an approach may encourage "myopic" views of the competitive arena. Firms which focus only on winning against immediate competitors may engage in actions which decrease their industry's value delivery, relative to competing offerings in the consumer market. Thus, a firm may generate a competitive strategy which does not achieve long term viability.

The premise argued in this paper has been that firms should expand the scope of their competitive strategy to include a perspective towards all four levels of competition, at market level, within the industry, between strategic groups, and within the strategic groups. This broader view, it is proposed, will help firms avoid the stagnating growth described in Levitt's myopia paper, as well as contribute to the longer term health and viability of larger value adding firms.

197

References

Ansoff, H. Igor, (1965), *Corporate Strategy*, New York: McGraw-Hill.

Barnes, J. H., (1984), "Cognitive Biases and Their Impact on Strategic Planning," *Strategic Management Journal*, 5, pp. 129-138.

Chandler, A., (1962), *Strategy and Structure*, Cambridge, MA: M.I.T. Press.

Daft, R. L., and K. E. Weick, (1984), "Toward a Model of Organizations as Interpretation Systems," *Academy of Management Review*, 9, pp. 284-295.

Day, George S., and Prakash Nedungadi, (1994), "Managerial Representations of Competitive Advantage," *Journal of Marketing*, 58(2, April), pp. 31-44.

Donaldson, G., and J. W. Lorsch, (1983), *Decision Making at the Top: The Shaping of Strategic Direction*, New York: Basic Books.

Fiol, C. Marlene, and Marjorie A. Lyles, (1985), "Organizational Learning," *Academy of Management Review*, 10(4), pp. 803-813.

Hambrick, D. C., (1983), "Some Tests of the Effectiveness and Functional Attributes of Miles and Snow's Strategic Types," *Academy of Management Journal*, 26, pp. 5-26.

Katz, D., and R. L. Kahn, (1966), *Social Psychology of Organizations*, New York: John Wiley & Sons.

Kiesler, S., and L. Sproull, (1982), "Managerial Response to Changing Environments: Perspectives on Problem Sensing From Social Cognition, *Administrative Science Quarterly*, 27, pp. 548-570.

Lawrence, P. R., and J. W. Lorsch, (1967), *Organization and Environment: Managing Differentiation and Integration*, Cambridge, MA: Harvard University Press.

Levitt, Theodore, (1975), "Marketing Myopia," *Harvard Business Review*, 53(5, September-October), pp. 26-48.

Porter, Michael E., (1980), *Competitive Strategy*, New York: The Free Press.

Smircich, L, and C. Stubbart, (1985), "Strategic Management in an Enacted World," *Academy of Management Review*, 10, pp. 724-736.

Stubbart, Charles I., (1989), "Managerial Cognition: A Missing Link in Strategic Management Research," *Journal of Management Studies*, 26(4), pp. 325-347.

Summers, E., (1980), *Strategic Behavior in Business and Government*, Boston: Little, Brown, Inc.

Thompson, J. D., (1967), *Organizations In Action*, New York: McGraw-Hill.

A GAME-THEORETIC ANALYSIS OF CAPACITY COMPETITION IN NON-DIFFERENTIATED OLIGOPOLISTIC MARKETS[1]

James A. Dearden, Gary L. Lilien and Eunsang Yoon

Lehigh University, USA

The Pennsylvania State University, USA

University of Massachusetts at Lowell, USA

Abstract

High capital investment industries often see regular cycles of over capacity followed by under capacity. We develop a game theoretic model and show that such cyclical behavior can exist in equilibrium, even if demand and prices are stable and if firms consider the capacity strategies of other firms. We discuss the implications of these findings for individual firm strategies that might reduce the impact of those cycles and for regulatory or industrial policies that might lead to more efficient market operations.

1. Introduction

The dynamics of high capital investment markets produce cycles of various sorts. Those cycles are highlighted in the business press, in numerous academic studies and in the everyday discussions with practicing managers. Consider the following:

> "Dennis H. Reilly of DuPont, speaking at a December 1993 briefing in London, pointed out that the major problem for the [Titanium Dioxide--TiO_2] industry is massive global overcapacity. ...Prices are at too low a level to justify investment, said Reilly. Although new investment is not needed now, when the recession ends and growth for TiO_2 picks up, there will be a shortage." (*Chemical & Engineering News*, January 3, 1994, p. 14.)

---

[1] The authors would like to acknowledge the contributions of Kalyan Chatterjee, Josh Eliashberg, John McNamara and Christophe Van den Bulte. This research was supported by Penn State's Institute for the Study of Business Markets

The dynamics of the TiO_2 industry are far from uncommon. The capacity cycle problem is summed up by:

> "In the familiar boom-bust pattern of the not-too-distant past, managers added production capacity, allowed overhead to swell, and stockpiled inventories in antipication of rising demand during expansions. When the economy tanked, they shut factories, laid off workers, iced new-product development, and purged excess inventories at distress prices." (*Fortune*, August 7, 1995, pp. 59-60.)

What is happening here? It appears that firms, privy to similar but noisy and often confusing information about what short-term and long-term demand for the output of their industry is or is likely to be, commit capital to add capacity or selectively delete capacity. In capital intensive industries, low capacity utilization adds a substantial cost burden to each unit sold; high capacity utilization leads to low unit costs and higher profits, as well as the tantalizing prospect of adding to firm profits by expanding.

In a monopoly, a firm might tune its production capacity to track demand cycles so that its capacity utilization was optimal in some long run profit maximizing sense, perhaps. Most oligopolies preclude such firm policies, however.

Is the phenomenon real or imagined? Consider Exhibits 1 to 3, based on aggregate US statistics on capacity utilization published every month by the Federal Reserve. The data show clear and persistent cycles in capacity utilization over the past several decades in the US. As the industry definition gets narrower, these cycles typically get more severe. Exhibit 1 shows that the capacity utilization of all US manufacturing, mining, and utilities combined has fluctuated from the low 70%'s to 90% over the last 30 years. As Exhibit 2 and 3 illustrate, similar patterns exist in individual manufacturing industries, but they are generally more severe and erratic, with fluctuations ranging from the low 40%'s to peaks of nearly 110%.

Are these just natural business cycles that promote a competitive marketplace? We think not: "...the structure of an industry may be so dysfunctional to the results of competition that collective action is appropriate to fix it. In such an instance the workings of the market produce neither efficiency nor profit" (Bower, 1986, p 14). And these fluctuations are not good for customers either, who see widely varying levels of supply assurance and prices. While these cycles may result from strategic competition for market leadership, they are dreaded by buyers, sellers and government regulators alike. Hence this paper will investigate some possible causes for these cycles and speculate what actions firms and regulators might take to address the situation.

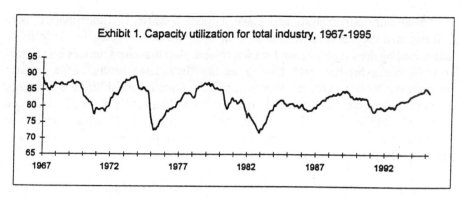

Exhibit 1. Capacity utilization for total industry, 1967-1995

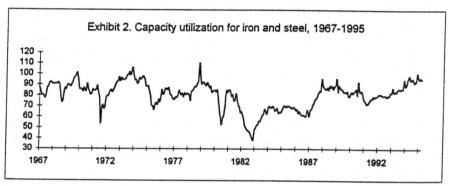

Exhibit 2. Capacity utilization for iron and steel, 1967-1995

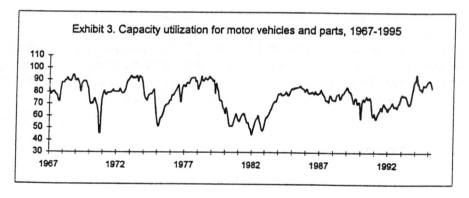

Exhibit 3. Capacity utilization for motor vehicles and parts, 1967-1995

To simplify our analysis and make it more specific, consider mature, non-differentiated oligopolistic industries. In such industries price is likely to be determined by the supply-demand market relationship that characterizes commodity or near-commodity markets. Due to the undifferentiated nature of the product, competition in such markets involves perceptions of quality, reliability and assurance of supply, elements that are closely related to a selling firm's share of production capacity. Many chemicals, metallic products, and electrical and electronic components are relevant examples. In such markets production cost is related to capacity utilization and market share is related to capacity share.

In the next section, we briefly review some of the related literature on the topic. Then, we describe the emprical findings in the Dearden, Lilien, and Yoon (1996) empirical analysis of the titanium dioxide (TiO_2) and Zicron industries. (Zicron is a fictitious name to preserve the proprietary nature of the data.) We use these empirical findings to aid in our choice of a game theoretic model to examine capacity cycles. Three findings, in particular, are relevant.

First, capacity cycles are often preceded by a market demand shock. Second, the empirical findings indicate that capacity, and not price, determines market shares. Hence, we choose a model of product-market competition whose equilibrium has this feature. Third, the exploratory analysis suggests that both pre-capacity and post-capacity marginal costs are constant in output.

In the following section we develop a game theoretic model incorporating many of these phenomena. The results of that model suggest that after a demand shock, the lack of capacity change coordination among an industry's firms prompts capacity cycles. We then present evidence that capacity changes in the titanium dioxide industry are consistent with the equilibrium results of the game theoretic model. In the final section, we discuss the implications and limitations of our work.

2. Related Literature

The literature in oligopoly theory related to our research has primarily considered capacity expansion decisions. Friedman (1983, Chapter 7), Gilbert (1986), and Fudenberg and Tirole (1986) give excellent surveys of different aspects of this literature. A more recent examination of capacity and competition is by Gal-Or (1994). The entry model of Dixit and Shapiro (1985) also shares some features of interest with the problem we consider.

The models of capacity expansion in this literature differ in several respects. First, some treat capacity as a physical limit on production, so that capacity expansion relaxes a constraint (for example, Prescott, 1973). Others treat additions to the capital stock as 'deepening' capital (or capacity) by shifting the marginal cost curve downwards (for example, Flaherty, 1980). Friedman (1983, p. 166) describes this a being "nearer to the spirit of neoclassical marginalist economics," and is the tack that we follow. (Note that we use the terms capacity and capital synonymously in this section.)

Second, the literature differs on whether a dominant firm exists. The seminal articles of Spence (1977) and Dixit (1980) and various sequels, such as Fudenberg and Tirole (1983), Ware (1984), and Arvan (1986) assume that there is a dominant firm, which moves first to add capital. Ghemawat (1984) shows that the firm with the greatest installed capacity (by assumption, the one with the lowest cost of additional investment) will add to capacity when only one firm is allowed to do so.

In the absence of 'natural' market leaders (like first entrants in models where firms enter sequentially over time), the presence of dominant firms cannot be assumed but must be endogenous to the analysis. (We will do this later by modifying the extensive form, without imposing the arbitrary constraint that only a single firm can add capacity in any given period.)

Third, the literature differs on whether or not capacity is continuous or "lumpy." We follow Ghemawat (1984) and much of the operations research literature (see, for example, Friedenfelds, 1981) in assuming that additions to capacity are 'lumpy' and that financial constraints may preclude firms from adding more than a given number of lumpy units of capital in a period.

Fourth, the literature differs in the consideration of production and demand dynamics. Dixit (1980), Ware (1984), and Arvan (1986) examine models in which firms compete in an output market in only one period. These models therefore cannot capture the overinvestment-underinvestment cycle that is endemic in many industries. Spence (1977), Friedman (1983, Chapter 7), Flaherty (1980), and Benoit and Krishna (1987), among others, on the other hand, consider dynamic models in which firms add capital and compete in an output market in an infinite number of periods. We will develop a two period model, the minimum number of periods needed to allow cycles to emerge.

3. Capacity Cycles in the TiO_2 and Zicron Industries

Dearden, Lilien, and Yoon (1996) analyze the capacity addition/deletion decisions for the TiO_2 and Zicron industries for the years 1970-1985. Exhibit 4 displays the results of this analysis qualitatively. Theier results generally suggest that overcapacity and undercapacity cycles are likely to occur because firms, acting on market signals,

Exhibit 4

Qualitative Summary of Empirical Results on the Dynamics of Capacity Addition and Deletion Decisions in the Titanium Dioxide and Zicron Industries

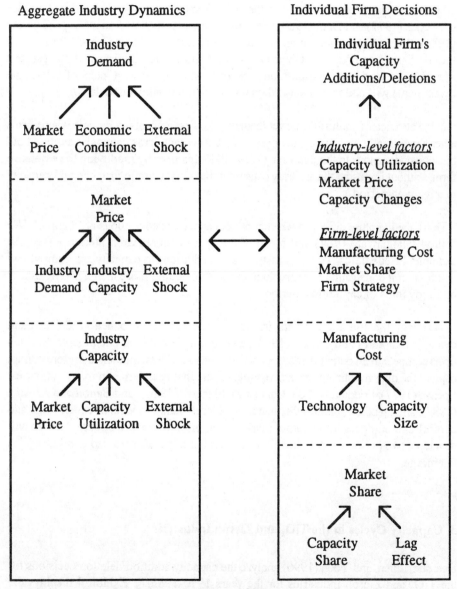

Source: Dearden, Lilien, and Yoon (1996)

simultaneously add (perhaps too much) capacity in good times, and delete (perhaps too much) capacity in poor times.

The empirical models they report fit and predict aggregate industry dynamics and individual firm decisions well. They found a high degree of sensitivity of a given firm's capacity addition and deletion decisions to other firms' capacity changes and industry capacity utilization. This finding confirms simultaneous, competitive firm behavior. They also found, via discriminant analyses, that a firm's price and its market share may be due to plant-specific factors or strategic differences, issues that we will explore in the next section and discuss later in more detail.

4. A Formal Model of Competitive Capacity Decisions

The empirical analysis in Dearden, Lilien, and Yoon (1996) suggests that firm capacity addition and deletion behavior can be explained by a series of regression models that do not incorporate knowledge or anticipation of competitive actions. Under such circumstances, where firms ignore the strategic actions of others, fluctuations in market demand lead to capacity cycles. But what if demand were stable over time and firms did consider the actions of other firms in their decision processes? Would we still see such cycles?

The results from Dearden, Lilien, and Yoon (1996) and the theoretical literature on capacity expansion (Section II) lead us to consider a model that should admit the following features:

i.   As suggested by the preliminary analysis, pre-capacity and post-capacity marginal costs are both constant in output.
ii. Because capacity, and not price, determines market shares, we model the product market competition as Cournot competition.
iii. Capital addition both relaxes a capacity constraint and lowers marginal cost for outputs above the previous capacity output and below the new capacity output.
iv. There is no dominant firm or 'natural' market leader.
v.   Capital addition is lumpy.
vi. There are at least two periods of product-market competition and potential capital change.

4.1. Model Formulation

Consider an industry with two firms, indexed as i=1,2. There are three important elements to our model: (i) the cost functions, (ii) the market demand function, and (iii) the timing of the firms' decisions.

The cost structure. Firm i's cost function has two components -- fixed and variable costs -- and both of these components depend on the firm's capital for capacity. Firm i's fixed cost function is $F_{it}(K_{i(t-1)}+d_{it})$; and its variable cost function is $V_{it}(K_{i(t-1)}+d_{it},Q_{it})$; where $K_{i(t-1)}$ is firm i's capital at time t-1, $d_{it} \in \{0,x\}$ is the capital that firm i changes in period t, and Q_{it} is firm i's output at time t. The capital change is constrained to be either 0 or x; that is, investment is lumpy.

We assume that $F_{it}(K_{i0}+x) > F_{it}(K_{i0})$, i.e., that capital addition raises fixed costs. In particular, we examine a specific form of the cost function:

$$F_{it}(K_{it})=\begin{cases} E_i & \text{if } K_{it} = K_{i0} \\ \overline{F}_i & \text{if } K_{it} = K_{i0}+x \end{cases} \tag{1}$$

where $\overline{F}_i > E_i$. We also assume that $V_{it}(K_{i0}+x,Q_{it}) \le V_{it}(K_{i0},Q_{it})$; capital addition for capacity does not raise variable costs. In particular, the variable cost function is

$$V_{it}(K_{it},Q_{it}) = \begin{cases} \underline{c}_i Q_{it} & \text{if } K_{it} = K_{i0} \text{ and } Q_{it} \le \hat{Q}_i \\ \underline{c}_i \hat{Q}_{it} + \overline{c}_i(Q_{it}-\hat{Q}_i) & \text{if } K_{it} = K_{i0} \text{ and } Q_{it} > \hat{Q}_i \\ \underline{c}_i Q_{it} & \text{if } K_{it} = K_{i0}+x \end{cases} \tag{2}$$

where $\hat{Q}_i$ denotes firm i's output at capacity, $\underline{c}_i$ denotes firm i's marginal cost for pre-capacity outputs, $Q_i \le \hat{Q}_i$, and $\overline{c}_i$ denotes firm i's marginal cost for for post-capacity outputs (where $\overline{c}_i > \underline{c}_i$). Thus, marginal cost is greater for pre-capacity outputs than for post-capacity outputs. Also, capital addition from K_{i0} to $K_{i0}+x$ increases capacity so that it is no longer binding at all equilibrium output rates. With the capital level $K_{i0}+x$, firm i then produces at marginal cost $\underline{c}_i$.

The market demand function. The firms produce an undifferentiated product, compete in a Cournot market, and face a linear demand curve

$$P_t = a - b(Q_{1t} + Q_{2t}) \tag{3}$$

where P_t denotes the industry price at time t, and $a, b > 0$ are parameters.

The extensive form and timing. We consider a 4-stage game, where t=1,2 correspond to the first period and t=3,4 correspond to period 2. In the first period: in stage 1, the firms simultaneously set capital levels; in stage 2, there is Cournot (quantity) competition. The second period is identical to the first period. Exhibit 5 presents the extensive form of this game.

The profit function. From the cost structure, market demand, and the extensive form, firm i's profit function is

$$\pi_i = [a - b(Q_{i2}, Q_{j2})]Q_{i2} - V_{i2}(K_{i2}, Q_{i2}) - F_{i2}(K_{i2}) \tag{4}$$

$$+ \delta \,[[a - b(Q_{i4}, Q_{j4})]Q_{i4} - V_{i4}(K_{i4}, Q_{i4}) - F_{i4}(K_{i4})\,],$$

where $\delta = 1/(1\text{-discount rate})$ denotes the discount factor from period 1 to 2.

4.3. Model Results

Existence of Equilibrium. Establishing the existence of a subgame perfect equilibrium is straightforward. Given capital stocks and our assumptions about the nature of competition, at stages 2 and 4, there is a well-defined strictly concave profit function for each firm. A unique Nash equilibrium exists for the stages 2 and 4 Cournot games. The crucial assumption for existence of a sub-game perfect equilibrium is that of finite action in stages 1 and 3 (capital decision stages). With a finite action space, an equilibrium exists by Nash's theorem.

One important characteristic of the equilibrium is that the periods are strategically independent. That is, the period-2 equilibrium play is independent of the period-1

Exhibit 5

The extensive form of period 1 of the capacity choice game

Period 1
Stage 1   Stage 2

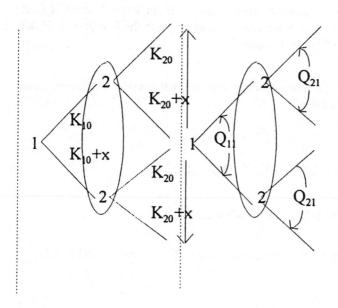

Note: The extensive form of period 2 of the capacity choice game is identical to the
extensive form of the game in period 1.

equilibrium play, and vice versa. (The argument is similar to Selten's chain store paradox story, see Selten, 1978.)

Characterizing the equilibrium. We first analyze the Cournot competition stages -- stages 2 and 4. Firm i's equilibrium output is

$$Q_{it}^* = \frac{a - 2c_i + c_j}{3b}, \tag{5}$$

where $c_i = \bar{c}_i$ if $Q_{it}^* > \hat{Q}_i$ and $c_i = \underline{c}_i$ if $Q_{it}^* \leq \hat{Q}_i$. An analogous condition holds for firm j. The equilibrium profit (induced by the pure-strategy equilibrium outputs) in stage t=2,4 for firm i is

$$\pi_{it} = \begin{cases} \dfrac{1}{9b}[a - 2\bar{c}_i + c_j]^2 + (\bar{c}_i - \underline{c}_i)\hat{Q}_i - E_i & \text{if } K_{it} = K_{i0} \\[2mm] \dfrac{1}{9b}[a - 2\underline{c}_i + c_j]^2 - \bar{F}_i & \text{if } K_{it} = K_{i0} + x. \end{cases} \tag{6}$$

given $c_j = \underline{c}_j, \bar{c}_j$.

Due to the simultaneous capital addition/deletion decisions, there are multiple equilibria in stages 1 and 3 -- two pure-strategy equilibria and one mixed-strategy equilibrium. This creates a difficulty when examining comparative statics. We then examine what we consider to be the most interesting and plausible equilibrium -- the mixed-strategy equilibrium.

The following two inequalities are sufficient for the stage-1 and stage-3 capacity choice games to be characterized as a battle-of-the-sexes game and hence for the existence of a mixed-strategy equilibrium. First, given that firm j chooses capital K_{j0}, we assume firm i earns greater profit by choosing capital $K_{i0} + x$ than by choosing K_{i0}. That is, the following inequality for the stage t (t=2,4) equilibrium profit holds:

$$\pi_{it}(K_{i0} + x, K_{j0}) - \pi_{it}(K_{i0}, K_{j0})$$
$$= (\frac{1}{9b}[a - 2\underline{c}_i + \bar{c}_j]^2 - \bar{F}_i) - (\frac{1}{9b}[a - 2\bar{c}_i + c_j]^2 + (\bar{c}_i - \underline{c}_i)\hat{Q}_i - E_i) > 0. \tag{7}$$

Second, given that firm j chooses capital $K_{j0}+x$, we assume firm i earns greater profit by choosing capital K_{i0}. That is, the following inequality for the stage t (t=2,4) equilibrium profit holds:

$$
\begin{aligned}
\pi_{it}(K_{i0}+x, K_{j0}+x) &- \pi_{it}(K_{i0}, K_{j0}+x) \\
= (\frac{1}{9b}[a-2\underline{c}_i+\underline{c}_j]^2 - \overline{F}_i) &- (\frac{1}{9b}[a-2\overline{c}_i+\underline{c}_j]^2 + (\overline{c}_i-\underline{c}_i)\hat{Q}_i - \underline{F}_i) < 0.
\end{aligned}
\tag{8}
$$

Given the demand and cost structure, Exhibit 6 lists firm i's stage-t (t=2,4) equilibrium profits as a function of the existing stage-t capacities (and hence as a function of the investments made before stage t).

We would like to note one important interpretation of the inequalities in expressions (7) and (8). The firms, prior to period 1, had the optimal capital stocks. Then, with a permanent increase in market demand at the onset of period 1, industry (i.e. joint) profit is maximized by the addition of one unit of capital. Hence, with this increase in demand and prior to any changes in capacity, the industry has undercapacity. As we demonstrate in this section, the permanent increase in market demand can generate undercapacity/overcapacity cycles.

We now analyze stages 1 and 3 assuming that the inequalities in expressions (7) and (8) hold. In stage 1, there are two pure strategies in which one firm chooses $K_{i0}+x$ and the other chooses K_{j0}. There is also the mixed strategy equilibrium in which both firms choose $K_{i0}+x$ with some probability. In the mixed strategy (and symmetric) equilibrium, the probability with which a firm chooses $K_{i0}+x$ depends on his opponent's payoff structure. Firm j randomizes between choosing capital $K_{j0}+x$ (with probability p_j) and capital K_{j0} (with probability $1-p_j$) so that firm i is just indifferent to choosing capital level K_{i0} and level $K_{i0}+x$. The mixed strategy equilibrium requires this indifference by firm i. Otherwise, if firm i did strictly better by say K_{i0}, then it would choose K_{i0} with probability 1 and not play a mixed strategy.

To calculate the equilibrium probability, p_j^*, that firm j chooses $K_{j0}+x$, we set firm i's expected profit from choosing capital $K_{i0}+x$ equal to its expected profit from choosing K_{i0}. These expected profits are determined by the equilibrium profits to the stage-2 Cournot game, and are stated in Exhibit 6. Given that firm j adds capital with probability p_j^*, firm i's expected profit from adding capital is

Exhibit 6

Firm i's state-t (t=2,4) equilibrium profit as a function of the state-t (t=2,4) capital stocks

		Firm j: $k_{jo} + x$	Firm j: k_{jo}
Firm i	$k_{io} + x$	$\frac{1}{9b}\left[a - 2\underline{c}_i + \underline{c}_j\right]^2 - \bar{F}_i$	$\frac{1}{9b}\left[a - 2\underline{c}_i + \bar{c}_j\right]^2 - \bar{F}_i$
Firm i	k_{io}	$\frac{1}{9b}\left[a - 2\bar{c}_i + \underline{c}_j\right]^2 + (\bar{c}_i - \underline{c}_i)\hat{Q}_i - \underline{F}_i$	$\frac{1}{9b}\left[a - 2\bar{c}_i + \underline{c}_j\right]^2 + (\bar{c}_i - \underline{c}_i)\hat{Q}_i - \underline{F}_i$

$$\pi_i(K_{i0}+x) = p_j^* \pi_i(K_{i0}+x, K_{j0}+x) + (1-p_j^*)\pi_i(K_{i0}+x, K_{j0}) \tag{9}$$

$$= p_j^* \left[\frac{1}{9b}\left[a - 2\underline{c}_i + \underline{c}_j\right]^2 - \overline{F}_i \right] + (1-p_j^*)\left[\frac{1}{9b}\left[a - 2\underline{c}_i + \underline{c}_j\right]^2 - \overline{F}_i\right].$$

and firm j's expected profit from not adding capital is

$$\pi_i(K_{i0}) = p_j^* \pi_i(K_{i0}, K_{j0}+x) + (1-p_j^*)\pi_i(K_{i0}, K_{j0}) \tag{10}$$

$$= p_j^* \left[\frac{1}{9b}\left[a - 2\overline{c}_i + \underline{c}_j\right]^2 + (\overline{c}_i - \underline{c}_i)\hat{Q}_i - \underline{F}_i \right]$$

$$+ (1-p_j^*)\left[\frac{1}{9b}\left[a - 2\overline{c}_i + \underline{c}_j\right]^2 + (\overline{c}_i - \underline{c}_i)\hat{Q}_i - \underline{F}_i\right].$$

Setting (9) = (10), i.e.,

$$p_j^* \pi_i(K_{i0}+x, K_{j0}+x) + (1-p_j^*)\pi_i(K_{i0}+x, K_{j0})$$

$$= p_j^* \pi_i(K_{i0}, K_{j0}+x) + (1-p_j^*)\pi_i(K_{i0}, K_{j0}),$$

and solving for p_j^*, yields

$$p_j^* = \frac{-9b(\overline{F}_i - \underline{F}_i) - 9b\hat{Q}_i(\overline{c}_i - \underline{c}_i) + 4a(\overline{c}_i - \underline{c}_i) + 4\overline{c}_j(\overline{c}_i - \underline{c}_i) - 4(\overline{c}_i^2 - \underline{c}_i^2)}{(\overline{c}_i - \underline{c}_i)(\overline{c}_j - \underline{c}_j)}. \tag{11}$$

We are interested in determining the likelihood of undercapacity/overcapacity cycles. In our model, the equilibrium strategies and payoffs in each period are identical. Therefore, suppose there is a "large" increase in demand and that each firm's capacity constraint imposes a large loss in profit. That is, each firm has a dominant strategy to choose capital $K_{i0}+x$ and thus increase capacity in period 1. If each firm adds capital in period 1 with probability 1, then each firm optimally chooses that capital level in period 2, and there are no capacity cycles in the industry. Similarly, suppose the capacity constraint is not binding in period one and each firm has a dominant strategy to keep the present capital K_{i0} and the associated capacity. If neither firm adds capital in period 1 with probability 1, then each firm optimally chooses that capital level in period 2, and there are again no capacity cycles in the industry. We therefore do not observe capacity cycles in our model when the firms' capacity strategies are deterministic. Rather, capacity cycles occur when the firms randomize between adding capacity and not adding capacity. The firms will randomize when the capacity constraint imposes a large enough loss in profit (so that the firms do not retain their present capacity with probability 1) and not too large of a loss is profit (so that the firms do not add capacity with probability 1). In this intermediate stage of profit loss associated with the capacity constraint, the firms play a battle-of-the-sexes game in the capacity addition stage, and we observe a mixed strategy equilibrium. The probability of observing a capacity cycle is

$$\text{prob(capacity cycle)} = \text{prob(both firms choose } K_{i0}+x \text{ in period 1)} \qquad (12)$$
$$\times \text{prob(both firms choose } K_{i0} \text{ in period 2)}$$
$$+ \text{prob(both firms choose } K_{i0} \text{ in period 1)}$$
$$\times \text{prob(both firms choose } K_{i0}+x \text{ in period 2)}.$$

Comparative Statics for the Mixed-Strategy Equilibrium. Now, we examine the effects of the cost structures on the probability of an overcapacity/undercapacity cycle. We first consider the effect of firm i's cost structure on the probability, in a mixed-strategy equilibrium, that firm j chooses $K_{j0}+x$ and adds capacity. With an increase in firm i's initial capacity output, $\hat{Q}_i$, it is relatively less profitable for firm i to choose capital $K_{i0}+x$. That is, with the initial capital K_{i0}, capacity is not as binding and firm i has a smaller incentive to remove this capacity constraint. For firm i to remain indifferent to K_{i0} and $K_{i0}+x$, firm j must choose $K_{j0}+x$ with a smaller probability. From (11), we derive

$$\frac{\partial p_j^*}{\partial \hat{Q}_i} = -\frac{9b}{4(\bar{c}_j - \underline{c}_j)} < 0. \qquad (13)$$

With an increase in firm i's post-capacity marginal cost $\bar{c}_i$, it is relatively more profitable for firm i to remove the capacity constraint by choosing capital $K_{i0}+x$. For firm i to remain indifferent to K_{i0} and $K_{i0}+x$, firm j must choose $K_{j0}+x$ with a greater probability. From (11), we derive

$$\frac{\partial p_j^*}{\partial \bar{c}_i} = \frac{-(\bar{c}_i^2 - \underline{c}_i^2) + 9b(\overline{F}_i - \underline{F}_i)}{4(\bar{c}_i - \underline{c}_i)^2(\bar{c}_j - \underline{c}_j)} > 0. \tag{14}$$

This term is positive because if $-(\bar{c}_i^2 - \underline{c}_i^2)/9b + (\overline{F}_i - \underline{F}_i) < 0$ firm i has a dominant strategy to choose $K_{i0}+x$ and add capacity. The net fixed cost of adding capacity is small compared to the net marginal cost gains of adding capacity.

Next, with an increase in firm i's pre-capacity marginal cost, $\underline{c}_i$, it is relatively less profitable for firm i to remove the capacity constraint by choosing capital $K_{i0}+x$. For firm i to remain indifferent to K_{i0} and $K_{i0}+x$, firm j must choose $K_{j0}+x$ with a smaller probability. From (11), we derive

$$\frac{\partial p_j^*}{\partial \underline{c}_i} = -\frac{(\bar{c}_i^2 - \underline{c}_i^2) + 9b(\overline{F}_i - \underline{F}_i)}{4(\bar{c}_i - \underline{c}_i)^2(\bar{c}_j - \underline{c}_j)} < 0. \tag{15}$$

We now consider the effect of firm j's cost structure on the probability, in a mixed-strategy equilibrium, that firm j chooses $K_{j0}+x$ and adds capacity. With an increase in firm j's post-capacity marginal cost $\bar{c}_j$, it is relatively more profitable for firm i to remove the capacity constraint by choosing capital $K_{i0}+x$. For firm i to remain indifferent to K_{i0} and $K_{i0}+x$, firm j must choose $K_{j0}+x$ with a greater probability. From (11), we derive

$$\frac{\partial p_j^*}{\partial \bar{c}_j} = -\frac{[(a - 2\bar{c}_i + \underline{c}_j)^2 - (a - 2\underline{c}_i + \underline{c}_j)^2] + 9b[(\bar{c}_i - \underline{c}_i)\hat{Q}_i + (\overline{F}_i - \underline{F}_i)]}{(\bar{c}_j - \underline{c}_j)^2(\bar{c}_i - \underline{c}_i)} > 0. \tag{16}$$

From the inequality in expression (8), i.e., part of our sufficient condition for a mixed-strategy equilibrium, (16) is positive.

With an increase in firm j's pre-capacity marginal cost $\underline{c}_j$, it is again relatively more profitable for firm i to remove the capacity constraint by choosing capital $K_{i0}+x$. For firm i to remain indifferent to K_{j0} and $K_{j0}+x$, firm j must choose $K_{j0}+x$ with a greater probability. From (11), we derive

$$\frac{\partial p_j^*}{\partial \underline{c}_j} = \frac{[(a-2\bar{c}_i+\bar{c}_j)^2-(a-2\underline{c}_i+\bar{c}_j)^2]+9b[(\bar{c}_i-\underline{c}_i)\hat{Q}_i+(\bar{F}_i-E_i)]}{(\bar{c}_j-\underline{c}_j)^2(\bar{c}_i-\underline{c}_i)} > 0. \quad (17)$$

From the inequality in expression (7), (17) is positive.

The Regions of the Mixed- and Pure-Strategy Equilibria. In Exhibit 7 we examine the effects of $(\underline{c}_j, \bar{c}_j)$ on the equilibrium strategy by firm j. In Region I_A, with c_j large and $\underline{c}_j$ small, firm j has a dominant strategy to choose $K_{j0}+x$ and increase capacity. In Region I_A, with $\bar{c}_j$ small and $\underline{c}_j$ large, firm j has a dominant strategy to choose K_{j0} and not add capacity. In Region III_A with intermediate values of $(\underline{c}_j, \bar{c}_j)$ and a mixed-strategy equilibrium (where firm i randomizes between choosing K_{j0} and $K_{j0}+x$), firm j randomizes between choosing K_{j0} and $K_{j0}+x$. To understand the region with the mixed strategies, consider the isoprobability line labeled $p_j^* = 0.5$. This line identifies all values of $\bar{c}_j$ and $\underline{c}_j$ for which firm j chooses K_{j0} and $K_{j0}+x$ each with probability 0.5. To derive this isoprobability line, we set the probability p_j^* as a function of $(\underline{c}_j, c_j)$ equal to 0.5. That is,

$$0.5 = p_j^*(\underline{c}_j, \bar{c}_j). \quad (18)$$

Taking the total differential of equation (18), we derive

$$0 = \frac{\partial p_j^*}{\partial \underline{c}_j}d\underline{c}_j + \frac{\partial p_j^*}{\partial \bar{c}_j}d\bar{c}_j. \quad (19)$$

Then, from (19), we derive

$$\frac{d\bar{c}_j}{d\underline{c}_j}\bigg|_{p_j^*=0.5} = -\frac{\partial p_j^*/\partial \underline{c}_j}{\partial p_j^*/\partial \bar{c}_j} < 0. \quad (20)$$

Exhibit 7

The effect of $\bar{c}_j$ and $\underline{c}_j$ on firm j's capacity decisions

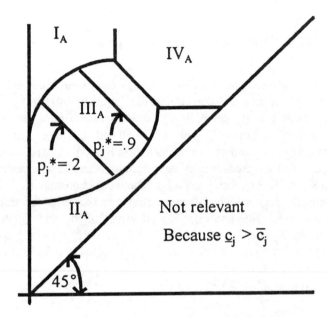

Marginal cost for post-capacity output, $\underline{c}_j$

I_A

IV_A

III_A

$p_j^*=.9$

$p_j^*=.2$

II_A

Not relevant

Because $\underline{c}_j > \bar{c}_j$

$45°$

Marginal cost for post-capacity output, $\bar{c}_j$

Notes:

Region I_A:	Firm j has a dominant strategy to choose $K_{j0}+x$.
Region II_A:	Firm j has a dominant strategy to choose K_{j0}.
Region III_A:	There is a mixed-strategy equilibrium. Two isoprobability lines are labeled.
Region IV_A:	Firm j exits.

Similarly, the slope of the each of the isoprobability lines is negative. Also, from expressions (7) and (8), p_j^* is increasing in $(\underline{c}_j, \bar{c}_j)$. In Region IV_A, costs are sufficiently large so that firm j exits the industry.

In Exhibit 8 we examine the effects of $(\hat{Q}_i, \bar{c}_i)$ on the equilibrium strategy by firm j. For this diagram, we consider the case for which (7) and (8) are satisfied, and hence firm j does not have a dominant strategy. In Region I_B, with $\hat{Q}_i$ small and c_i large, firm i has a dominant strategy to choose $K_{i0}+x$ and add capacity. Firm j optimally responds to i's choice of $K_{i0}+x$ by choosing K_{j0}. In Region II_B, with $\hat{Q}_i$ large and c_i small, firm i has a dominant strategy to choose K_{i0} and not add capacity. Firm j optimally responds by choosing $K_{j0}+x$ and adding capacity. In Region III_B, with intermediate values of $\bar{c}_i/\hat{Q}_i$, there is a mixed-strategy equilibrium. The slope of an isoprobability line is

$$\left. \frac{d\bar{c}_j}{d\hat{Q}_i} \right|_{p_j^* \text{ constant}} = - \frac{\partial p_j^*/\partial \hat{Q}_i}{\partial p_j^*/\partial \bar{c}_j} > 0. \tag{21}$$

Moreover, p_j^* is increasing in $\bar{c}_i$ and decreasing in $\hat{Q}_i$.

In Exhibit 9 we examine the effects of $(\hat{Q}_i, \underline{c}_i)$ on the equilibrium strategy of firm j. For this diagram, we again consider the case for which (7) and (8) hold, and hence firm j does not have a dominant strategy. In Region I_C, with both $\hat{Q}_i$ and $\underline{c}_i$ large, firm i has a dominant strategy to choose $K_{i0}+x$ and add capacity. Firm j optimally responds to i's choice of $K_{i0}+x$ by choosing K_{j0}. In Region II_C, with both $\hat{Q}_i$ and $\bar{c}_i$ small, firm i has a dominant strategy to choose K_{i0} and not add capacity. Firm j optimally responds by choosing $K_{j0}+x$ and adding capacity. In Region III_C, with intermediate values of $\bar{c}_j + \hat{Q}_i$, there is a mixed-strategy equilibrium. The slope of an isoprobability line is

$$\left. \frac{d\underline{c}_j}{d\hat{Q}_i} \right|_{p_j^* \text{ constant}} = - \frac{\partial p_j^*/\partial \hat{Q}_i}{\partial p_j^*/\partial \underline{c}_j} < 0. \tag{22}$$

Moreover, p_j^* is decreasing in $(\bar{c}_i, \hat{Q}_i)$.

Exhibit 8

The effect of $(\hat{Q}_i, \bar{c}_i)$ on firm j's strategy.

Marginal cost for
pre-capacity
output, $\bar{c}_i$

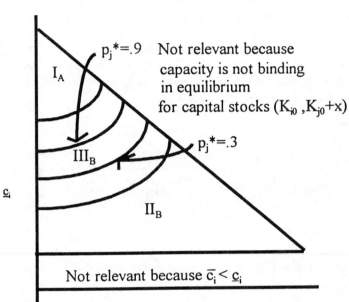

I_A

$p_j^*=.9$ Not relevant because
capacity is not binding
in equilibrium
for capital stocks $(K_{i0}, K_{j0}+x)$

$p_j^*=.3$

III_B

$\underline{c}_i$

II_B

Not relevant because $\bar{c}_i < \underline{c}_i$

Production capacity, $\hat{Q}_i$

Notes:

This figure considers the case in which firm j does not have a dominant strategy.

Region I_B:	Firm i has a dominant strategy to choose $K_{i0}+x$. Firm j best responds by choosing K_{j0}.
Region II_B:	Firm i has a dominant strategy to choose K_{i0}. Firm j best responds by choosing $K_{j0}+x$.
Region III_B:	Mixed-strategy equilibrium region. Representative isoprobability lines are labeled.

Exhibit 9

The effect of $(\hat{Q}_i, \underline{c}_i)$ on firm j's strategy

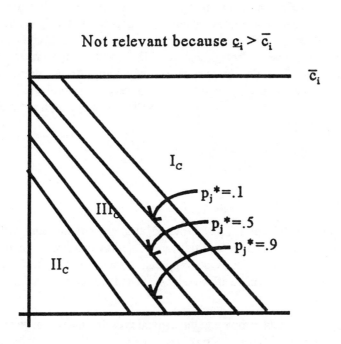

Marginal cost for post-capacity output, $\underline{c}_i$

Not relevant because $\underline{c}_i > \bar{c}_i$

$\bar{c}_i$

I_c

$p_j^* = .1$

$p_j^* = .5$

$p_j^* = .9$

III_c

II_c

Production capacity, $\hat{Q}_i$

Notes:

This figure describes the case for which firm j does not have a dominant strategy.

Region I_c: Firm i has a dominant strategy to choose K_{i0}. Firm j best responds by choosing $K_{j0}+x$.

Region II_c: Firm i has a dominant strategy to choose $K_{i0}+x$. Firm j best responds by choosing K_{j0}.

Region III_c: Mixed-strategy equilibrium region. Representative isoprobability lines are labeled.

To further understand the different regions of the equilibrium in Exhibit 9, fix $\hat{Q}_i$ and increase $\underline{c}_i$ so that the equilibrium moves from region II_C to region III_C. With this movement, there is a discontintuity in firm j's equilibrium strategy. In region II_C, firm i has a dominant strategy to choose capital K_{i0} +x and add capacity. Firm j best-responds by choosing capital K_{j0} and not adding capacity. With an increase in $\underline{c}_i$ and the movement to slightly above the boundary of regions II_C and III_C. Now, firm i no longer has a dominant strategy to add capacity. However, $\underline{c}_i$ is small enough so that firm i has a strong incentive to choose capital K_{i0}+x and add capacity. For firm i to remain indifferent between between adding and not adding capacity, firm j must add capacity with a large probability. Thus, with the movement from region II_C to region III_C, firm j's equilibrium strategy jumps from adding capacity with probability zero to adding capacity with a very large probability. There is an analogous discontintuity in firm i's equilibrium strategy with the movement from region III_C to region I_C.

4.3. Model Assessment

Our theoretical model admits a mixed strategy equilibrium, with both firms randomizing their capacity addition and deletion decisions, yielding a positive probability of capacity cycles. Our comparative statistics show how the likelihood of such cycles in theory can be affected by costs and capacity utilization.

How do these results compare to the dynamics of the markets in general or to those studied in Dearden, Lilien and Yoon (1996), TiO_2 in particular? Consider Exhibit 10, which suggests several behaviors that are consistent with our model.

Observation 1:   We see that individual firms' *simultaneous* capacity additions and maintenance (e.g., 1973-74 or 1982-83) are all followed by their *simultaneous* capacity deletions or maintenance (e.g., 1976-76 or 1984-85).

Observation 2:   The cycles (i.e., additions and deletions) of total industry capacity have occurred in parallel with the cycles (i.e., Highs and Lows) of *capacity utilization* with a typical lag of 1-2 years.

Observation 3:   Low and medium *cost* firms (e.g., A or B) have typically responded to industry fluctuations by their capacity cycles of additions and maintenance, while *high* cost firms (e.g., C and D) have responded to industry fluctuations by their capacity cycles of additions, *deletions*, and maintenance.

To generate an additional observation we focus on firms A and C, the "high" market share firms who mainly dictate the evolution of the market. If we assume that A and C are playing fixed strategies, then we would expect to see certain deterministic patterns of simultaneous moves dominate. (Both add, both delete, A adds, C doesn't, etc.) If we code the data as follows:

1. A+2 = A adds, C deletes [3]

2. A+1 = $\begin{cases} \text{A adds, C no change; or} \\ \text{A no change ,C deletes} \end{cases}$ [6]

3. A = $\begin{cases} \text{A adds, C deletes; or} \\ \text{A no change, C no change; or} \\ \text{A deletes, C adds} \end{cases}$ [6]

4. A-1 = $\begin{cases} \text{A no change, C adds; or} \\ \text{A deletes, C no change} \end{cases}$ [1]

5. A-2 = A deletes, C adds [0]

we have five possible "deterministic strategy" patterns relating A's and C's capacity strategies. For a deterministic strategy to dominate, we should expect one of these patterns to prevail.

The numbers in [ ] show the frequency with which the noted pattern occurs in Exhibit 10. Not surprisingly, the A - 1 and A - 2 patterns were virtually nonexistent (since A is the low cost firm). However, the high incidence of all three of the other patterns (3, 6 and 6) suggest capacity decisions that look like mixed strategies, providing qualitative support for our model.

5. Discussion and Conclusions

The dynamics of markets appear to lead to cycles of overcapacity and undercapacity. In this paper we have explored some of the forces that lead to those cycles.

Exhibit 10
Capacity Additions and Deletions* in the TiO$_2$ Industry

Year	Individual firm's capacity						aggregate industry capacity	capacity utilization**	
	A	B	C	D	E	F		1-yr lag	2-yr lag
1970	+	+	o	+	+	+	+	H	M
71	o	o	o	-	o	+	+	M	H
72	+	o	-	+	o	+	-	M	M
73	+	o	-	+	o	+	+	H	M
74	+	+	+	o	+	o	+	H	H
75	o	o	o	o	+	o	+	M	H
76	+	o	-	o	o	o	+	L	M
77	o	o	-	o	+	o	-	L	L
78	o	o	-	o	o	o	-	L	L
79	+	o	o	-	o	o	-	M	L
80	+	o	o	o	o	o	+	M	M
81	-	o	-	-	+	+	+	M	M
82	o	+	o	+	o	+	+	M	M
83	o	+	+	+	+	+	+	L	M
84	o	o	-	o	+	o	-	M	L
85	o	o	o	o	o	o	+	H	M

Individual Firm's Position

(a)Production cost***

    L L H H M M

(b)Market share****

    H L H M M L

* +: addition, -: deletion, and o: no change.
** H: 90% or higher, M: 80-90%, and L: 80% or lower.
*** H: relatively high, M: medium, and L: relatively low.
**** H: relatively high, M: medium, and L: relatively low.

These cycles are inefficient, whatever their causes, disrupting buyers, sellers and often leaving ultimate consumers with either higher prices or delayed access to products desired. What can sellers do to address these issues? Bower (1986, p. 221) suggests: "...it would be extremely helpful...if industry associations were asked to produce long term forecasts of supply/demand balance." What Bower suggests is that reducing uncertainty about the nature of demand would help coordinate capacity planning efforts. This will surely help, but our game-theoretic results indicate that unless individual company plans are coordinated, we are unlikely to see a cure.

Are there alternatives? In Japan, MITI helps coordinate the strategic plans of competitive companies. Firms give up some independence (legally in Japan, at least) in exchange for the benefits of coordination or consensus building. The stabilizing effects of such cartel-like coordination procedures reduce cyclical behavior but may do so at the cost of keeping inefficient producers in the market (Shaw and Shaw, 1983).

In the absence of the possibility of such coordination in many of Western industrialized nations, firms need to adopt other strategies. This paper serves as further support for the risks of operating in such markets: overcapacity/ undercapacity cycles are almost destined to occur. Flexible manufacturing systems, sharing or reallocating production capacity with other, countercyclical (or uncorrelated, at least) products can help, though (Breshnahan and Ramey, 1993). In addition, firms may need to budget for higher expected returns to deal with the risks inherent in operating in such industries.

Firms can take advantage of these situations. Although our models have not addressed the issue, there is clearly value in information about demand and likely competitive actions. Better industry and competitive intelligence is likely to pay large dividends for firms operating in such markets, signing longer terms supply contracts with customers during the onset of undercapacity, for example. We also speculate that there may be advantages to various forms of bluffing (announcing expansion/deletion plans for strategic rather than operational reasons).

There may well be regulatory implications here. If there were agreement on the likely level of demand, one could envision a situation where the government entertained bids for the rights to add new capacity and limited the amount of capacity to some multiple of the industry's estimate of the increase in demand. There are clearly many problems with the development of such a system, as the US government's recent experience with cellular phone franchises and bandwidth auctions has shown, but the idea may have some merit.

Our model and analysis and the above speculations have been exploratory. We have investigated only a simple model here (see Dearden, Lilien, and Yoon, 1996, for others) and one can envision many other causes of the capacity cycle phenomenon. We do not believe that this phenomenon has a single cause or set of causes; rather we believe that it would be valuable, in future research, to see how general the phenomenon is and to generate and develop a taxonomy of causes and possible cures. We hope that we have shed some light on some of the possible causes and that further work helps deepen that understanding.

References

Arvan, L. (1986), "Sunk Capacity Costs, Long-Run Fixed Costs and Entry Deterrence under Complete and Incomplete Information," *Rand Journal of Economics,* 17, 105-121.

Benoit, J.-P., and V. Krishna (1987) "Dynamic Duopoly: Prices and Quantities," *Review of Economic Studies,* 54, 23-35.

Bower, J. L. (1986), *When Markets Quake,* Boston, MA: Harvard Business School Press.

Dearden, J.A., G.L. Lilien and E. Yoon (1996), *Capacity Competition in Non-Differentiated Oligopolistic Markets,* The Pennsylvania State University, ISBM Report 1-1996.

Dixit, A. (1980), "The Role of Investment in Entry-Deterrence," *The Economic Journal,* 90, 95-106.

Dixit, A., and C. Shapiro (1985), "Entry Dynamics with Mixed Strategies," Working Paper, Princeton University.

Flaherty, M. T. (1980), "Industry Structure and Cost-Reducing Investment," *Econometrica,* 48, 1187-1209.

Friedenfelds, J. (1981), *Capacity Expansion Analysis of Simple Models with Applications,* New York: North-Holland.

Friedman, J. (1983), *Oligopoly Theory,* Cambridge: Cambridge University Press.

Fudenberg, D., and J. Tirole (1983), "Capital as Commitment: Strategic Investment to Deter Mobility," *Journal of Economic Theory,* 31, 227-250.

Fudenberg, D., and J. Tirole (1986), *Dynamic Models of Oligopoly,* Harwood.

Gal-Or, E. (1994), "Excessive Investment in Hospital Capacities," *Journal of Economics and Management Strategy,* 3, 53-70.

Ghemawat, P. (1984), "Capacity Expansion in the Titanium Dioxide Industry," *The Journal of Industrial Economics,* 33, 145-163.

Gilbert, R. (1986), "Preemptive Competition," in F. Mathewson and J. Stiglitz, eds., *New Developments in the Analysis of Market Structure,* Cambridge, MA: MIT Press.

Prescott, E. (1973), "Market Structure and Monopoly Profits: A Dynamic Theory, *Journal of Economic Theory,* 6, 546-557.

Selten, R. (1978), "The Chain-Store Paradox," *Theory and Decision*, 9, 127-159.

Shaw, R.W., and S.A. Shaw (1983), "Excess Capacity and Rationalization in the West European Synthetic Fiber Industry," *The Journal of Industrial Economics*, 32, 149-166.

Spence, A. M. (1977), "Entry, Capacity Investment, and Oligopolistic Pricing," *Bell Journal of Economics*, 8, 534-544.

Ware, R. (1984), "Sunk Costs and Strategic Commitment: A proposed Three-Stage Equilibrium," *The Economic Journal*, 94, 370-378.

INTEGRATING ADVERTISING AND PROMOTION WITH
THE ORGANIZATION'S "NONMARKETING" ACTIVITIES:
DYNAMIC CONCEPTS AND A COMPUTER-ASSISTED
    PROFIT/COST PLANNING APPROACH

by Morton I. Jaffe*

Abstract

Decisions on the volume, type, and content of
advertising and promotion usually result in
certain increased costs/investments/operating-
inefficiencies for other departments/functions
of the organization.   Therefore, marketing
planning must take these effects into account
(as well as the related resistances of the
managements of these departments to these
advertising and promotion plans).   Trends in
management teamwork, decentralization, re-
engineering, information technology, and
channel-relations strategy make conflict-
anticipating/reducing planning methods
particularly desirable.
    The marketing methods which tend to have
relatively greater adverse effects on the costs
and efficiency of "nonmarketing"
departments/functions include those that cause
demand to be irregular, unpredicted, or
otherwise unusual.   Favored methods include
those that increase demand in a regular,
predictable way.
    A computer-assisted, dynamic, interactive
model and method is provided for planning
advertising and promotion which includes the

* Baruch College, City Univ. of New York, USA

interdepartmental/interfunctional
cost/efficiency aspects of the planning process
which are usually omitted from other models and
methods.

A simulated case example is cited.

The application of selected game theory
concepts provides a helpful perspective.

Keywords
advertising, promotion, marketing
communications, game theory, conflict,
organization, group decision, interdepartmental
relations, planning, computer-methods.

1. Introduction

The process of planning advertising and sales
promotion, as integrated with other marketing
communications such as personal selling,
publicity, and public relations (Schultz 1992,
Belch & Belch 1995), usually involves
considering alternative strategies and plans for
the particular planning horizon (often one year)
including projecting the effects of the plan on
sales, gross profit, net profit, investment,
free cash flow, etc. This process has been
previously modeled using a computer-assisted
planning method by Jaffe (1985) building on a
model developed at Stanford University
(Eskin/Stanford U. 1970, Stanford U. 1973).

The advertising and promotion planning
process must increasingly take into account the
effects of alternative advertising and promotion
plans on various other constituencies of the
organization including (a) other departments
(i.e., manufacturing, finance, accounting,
etc.), (b) other functions that are often
considered partly marketing and partly
"nonmarketing" (e.g., inventory management,
transportation, other logistics, customer
servicing, purchasing), and (c), other
organizations (i.e., channel-of-distribution

members and suppliers) (Applegate/Harvard U. 1992 & 1993, Wal-Mart/Harvard U. 1994). As a result, models of this process need to be changed correspondingly to provide for these interactions and to facilitate the full planning process. In addition, assistance in understanding and dealing with the interactions and conflicts among members of planning groups can be obtained from game theory.

Thus, one thesis of this paper is the presentation of a more advanced computer-assisted planning model and method. Another thesis is the need for more recognition by marketing management of the needs and developments in this area.

In this paper, the term "nonmarketing" will be used as a composite for the departments and functions of the organization that are either (a) usually clearly separate from marketing (i.e., manufacturing, finance, accounting), and/or (b), usually considered to be partly marketing and partly "not marketing" (i..e, inventory management, transportation, other logistics, credit, order processing, other servicing, purchasing, other operations).

To further specify the terminology in this paper, the terms "advertising and promotion" and "marketing communications" will be used interchangeably, although readers will recognize that, based on developing usage, both "marketing communications" and "promotion" usually include advertising, sales promotion, personal selling, direct marketing, publicity, and public relations (Schultz 1992, Belch & Belch 1995), readers will also recognize that some people use the term "promotion" to mean "sales promotion" only.

It should also be noted that there is a suggestion, based on earlier studies, that the interdepartmental conflicts of the marketing department are substantial and the highest of any department in the organization (Walton, Dutton, & Cafferty 1969, & Walton & Dutton 1969. In addition, there are often differences and disagreements between marketing and "nonmarketing" that are not the result of the volume, type, and timing of advertising and promotion (the focus of this paper).

2. Perspectives on the Process of Planning Advertising and Promotion

Decisions on the volume, type and content of marketing communications (advertising and promotion) are likely to result in increased costs, decreased efficiency, and/or increased investment for many of the organization's so-called "nonmarketing" functions and departments. This is because, as will be described later, advertising and promotion often result in "nonmarketing" experiencing irregular, unpredicted, and other special demands which are almost certain to be costly, as well as, difficult to meet. While decreased costs and increased efficiency in these "nonmarketing activities" also may result (usually because of increased sales and scale of activity), this favorable outcome is somewhat more uncertain and is more likely to occur only longer term. As a result, the departments/people in charge of these "nonmarketing" functions tend to resist certain types of advertising and promotion plans or, at least, they try to get plans modified. Thus, it is important for advertising and promotion planning to integrate with "nonmarketing" in the development and negotiation of alternative marketing plans and decisions because the related increased costs and decreased efficiency are such important concerns (Jaffe 1995, Applegate/Harvard U. 1992, ibid. 1993, Wal-Mart/Harvard 1994). In the simulated case situation described later in this paper, the "nonmarketing" cost increases were 2.8% of cost of goods sold and the effect was to reduce projected profit by 9.7%.

From a related perspective, the advertising and promotion planning process can be usefully viewed as a dynamic, group, "cooperative" game in which the managers of the "nonmarketing" functions resist or otherwise seek to influence the advertising and promotion program because to the extent that this program is apt to result in inefficiencies or increased costs or investments for their own functions, while marketing management seeks to manage the problem and to have their plans approved and smoothly facilitated.

3. An Overview of the Model and Method

To better deal with the above problems, marketing management needs to give explicit attention to the nonmarketing costs, inefficiencies, and investment in its own process of planning the marketing program that it will propose to higher management or to a planning committee. By doing this, the marketing plan has better prospects for adoption and success and the decision process and its implementation functions much more smoothly. In effect, the proposed planning method in this paper provides in advance for the nonmarketing concerns before the marketing planning process gets too far.

To provide for the interactions and viewpoints involved, the computer-assisted planning model in this paper first forecasts the levels and timing of sales/demand depending on the marketing variables planned (i.e., levels, types, and timing of advertising, promotion, and price) using the sales-response elasticities which the planner or model specifies. The model also requires the planner to specifically adjust costs and profit margins for any "nonmarketing" inefficiencies (as well as efficiencies) resulting. (e.g., for changes in inventories, manufacturing costs, raw materials costs, transportation costs, financing, servicing, etc.). Without these adjustments that explicitly and quantitatively take into account the increases and decreases which the advertising and promotion plan may cause for the "nonmarketing" costs, the best (most profitable) marketing plan cannot be derived. Thus, a more advanced type of "what if" analyses can be made.

One effect of this approach is to caste in a new light the dynamic interfunctional, interdepartmental interactions involved and to facilitate making the marketing and nonmarketing functions more integrated rather than competitive.

The adverse cost/profit impact on the organizations "nonmarketing" of a marketing communications program has been shown to increase the Cost of Goods Sold and Other Expense by as much as one to five per cent of cost of goods sold and to decrease profits by as much as 10% or more.

Thus, some advertising/promotion strategies (budget, type, content, timing) have been found to be comparatively disadvantageous because they tend to increase these nonmarketing costs (aside from the direct costs of the marketing actions themselves and their effects on other marketing areas). To generalize, the activities which tend to be particularly disadvantageous to "nonmarketing" functions include short term price promotions, "blockbuster" promotions, promotions requiring product purchase, direct-action advertising, and the timing of promotions at high-sales-volume periods. Methods which tend to not increase these costs, or even decrease them, include brand-equity/image advertising, events marketing, smaller-scale and spaced promotions, and activities not scheduled at peak buying seasons.

Structurally, as noted, the model in this paper adds the interdepartmental and interfunctional dimension to a prior computer-assisted model (Jaffe 1985, Eskin/Stanford U. 1970 & 1973), the former including allowing the model operator to simulate interactively alternative levels of demand, advertising, promotion, and price, along with response-elasticity coefficients about the sales effects of these inputs.

Thus, an updated computer-assisted, interactive, descriptive planning model

(including diskette) is now provided for
planning month-by-month for one year (with a
version under development to provide for
planning year by year for five years).

4. Perspectives on Management Practices and
   Trends

A number of real world developments and trends
provide evidence of the need and direction for
models and methods which assist marketing
planning in integrating and handling conflicts
with "nonmarketing" considerations. Among these
developments are the recent and converging
trends in management teamwork, decentralization,
re-engineering and cost control, information
technology, and channel-relationship strategy
which make marketing planning methods that
better integrate and reduce conflict more
desirable and feasible. Among the specific
developments, trends, and needs are the
following:

4.1 The ongoing trends toward decentralized
decision-making, just-in-time inventory and
manufacturing, information technology,
relationship marketing, and channel-effect
models and methods, require that marketing
planning be more attuned to its effect on
"nonmarketing" costs and conflicts.

4.1.1 For example, the Frito-Lay Company
(Applegate/Harvard U. 1992, ibid. 1993) has
implemented a decentralized planning and
decision-making procedure whereby marketing
plans are proposed to, and must be approved by,
a multifunctional committee in each geographical
market area. In this process, the impacts of
marketing actions on "nonmarketing"
(manufacturing, purchasing, inventories,
transportation, finance, account, other
operations, etc.) are explicitly considered,
with the bonus/compensation of all concerned
dependent on area profit. In this context, the

effect of proposed marketing plans on the costs of "nonmarketing" are given explicit consideration, and directly affect the marketing manager's compensation, as well as directly affecting the compensation of the managements of other departments. Very significantly, the nonmarketing managers have been found to be less likely to oppose marketing actions that increase the costs or decrease the efficiency of their own departments/functions if these costs are explicitly considered in planning and if the results are shown to be favorable to the overall area profit plan. This is because their compensation will be based on the overall area profit.

4.1.2 Marketers like Coca Cola and Wal-Mart have implemented "channel-effect and profitability" models and methods which take into account, calculate, and prove the (favorable) effect of the manufacturer's advertising and promotion plan on the retailer or other channel intermediaries. These models focus on profit or free cash flow and build on both the marketing and nonmarketing relationships between the manufacturer and the retailer/channel (Jaffe 1995, Wal-Mart/Harvard U. 1994).

4.2 In the food processing industry, "based on a joint study, manufacturers and retailers are preparing a drive to slash $30 billion in costs over the next five years from the process of making, distributing, and selling food. Many grocers and wholesalers now stockpile goods when food makers offer discounts, leading to inefficient peaks and valleys in production. Through a program called Efficient Customer Response, in which 11 industry trade groups are participating, the processors aim to smooth out orders and, therefore, manufacturing cycles. A pioneer in such efforts, Quaker Oats Co. now pays promotional dollars regularly instead of largely at the end of each quarter." (Business Week 1/10/94).

4.3 It is well known that a number of prominent marketers have recently cut back on trade promotion in favor of brand-equity advertising and lower every-day prices. While the primary reason for this is not the "nonmarketing" costs, they are a factor.

5. How Marketing Communications Specifically Affects Nonmarketing Costs and Efficiency

The following is an overview of how marketing communications (advertising and promotion) tend to affect "nonmarketing" costs. Focus is on the effect of a one-year marketing plan and on the situation of a packaged-cereal marketer, with specifics for the illustration provided later.

5.1 Manufacturing costs (per unit) - most manufacturing is most efficient (low cost, high quality, meets shipment requirements and schedules, etc.) if the same amount of an unchanged product is produced year around. Also, adjustments for any seasonal effects and other special considerations are more efficient to the extent that they are regular and predictable. Thus, for example, if the volume or type of marketing communications causes/leads-to irregularities or surprises in the volume of production, requirements for special product/package types for promotion, etc., the per-unit costs of manufacturing the product may well increase (because of such factors as labor overtime, use of less efficient equipment and personnel, reduced maintenance on equipment, required capital investment, etc.). However, manufacturing cost (per unit) is also often favorably influenced by high volume, as long as the volume is not so high as to be outbalanced by the cost-increase or investment considerations just cited. Thus, company higher-management and manufacturing management are concerned with how marketing plan specifics may or may not change the nature and timing of demand, and therefore, change the cost of manufacturing. The accuracy of sales forecasts is an important factor which is, of course, affected by the level, type, and timing of advertising and promotion.

5.2 Costs of raw materials and supplies (per unit) are likely to be reduced by marketing communications because of any ability to qualify for higher discounts in purchasing, unless this cannot be planned in advance so that distress purchases, higher inventories, etc. are needed.

5.3 Inventory costs include costs of buying, receiving, handling, storing, insuring, financing, etc. The nonfixed portion of these costs change with demand.

5.4 Other Logistics Costs - On a per-unit basis, there may be a savings in logistics costs for transportation to customers, physical handling, etc. as a result of demand increasing more than variable logistics costs. However, there may be no savings if costs of transportation are strictly per-unit with no volume savings at the demand levels experienced.

5.5 Financial cost includes the cost of financing net accounts receivable, investment, and any bad debts. The net accounts receivable is the accounts receivable less the accounts payable to cover the added dollar sales/purchases due to promotion, advertising, etc. Some costs of financing tend to be a fixed cost, not a per-unit cost.

5.6 Other variable costs - this category is a catch-all in the model and includes costs of the following, if changed as a result of the marketing communications plan:
   billing (invoicing, receiving payment, inquiries).
   checking compliance with promotion terms.
   logistics of arranging advertising.
   purchasing (cost of purchasing increased raw

    materials and supplies)
    personnel efficiency, motivation, and
        satisfaction (if quantifiable)

5.7.   Overhead consists of the costs of
administration, selling expense, and other fixed
marketing and administrative expense.   It is
usually a fixed cost, not a per unit cost.

5.8.   Advertising Expenditure and Promotion
Expenditure   -   It should be noted that the
direct costs of the marketing communications
program are included in the expenditure/budget
of the promotion.   These are divided into
separate expenditures for advertising and for
promotion.
    Later in this paper, Table 2 indicates the
adjustments to "nonmarketing" costs projected to
result from the proposed advertising and
promotion plan costs used in the cited simulated
case example of a cereal marketer, resulting in
an estimated projected increase in nonmarketing
costs (due to the company's planned marketing
communications program for 1994) of over $1.8
million, 2.8% of Cost of Goods Sold and 9.7% of
projected profit.   (This cost increase and
reduction in the projected profit increase would
not derail the advertising and promotion plan
since it was still projected to directly
increase profits and since it would be useful
for other reasons, i.e., influencing wholesalers
and retailer support, maintaining market share
and SKUs, etc.).

6. Perspectives on the Relevance and Role of
   Game Theory

6.1 The concepts of game theory seem relevant
to the planning situation described in this
paper because of the following:

a.   It is a group decision among marketing and
"nonmarketing", involving usually four to eight
different individuals/perspectives, plus some

subgroups, plus an overall arbiter if needed
(Fudenberg & Tirold 1991, Hammond 1993).

b. There are frequent opinion differences and
conflicts which are resolved by negotiation and,
if necessary, by the ultimate arbiter
(Applegate/Harvard U. 1992, ibid., 1993, Moulin
1988).

c. It is a usually a non-zero-sum payoff
situation for each player. However, the payoffs
to each player differ and the narrowly-defined
payoffs may even be negative for some players.
The payoffs are also usually uncertain (Myerson
1991).

d. Each player can usually observe/determine
many, but not all, of the other player's actions
and only some of their consequences (Radner,
Myerson & Raskin 1986, Forges 1993).

e. Future expected benefits from
cooperation/lack-of-cooperation stimulate
cooperation in the current period. In this,
there may be some stability of rules and the
influence of corporate culture (Cremer 1986).

6.2 At present, the planning situation is so
complex that clear-cut modeling is unfeasible
without further investigation and experience.
The present difficulty in modeling, determining
equilibria or optimal solutions, and otherwise
applying gaming principles to the planning
situation include the following:

a. The payoffs to each player are unequal,
difficult to determine, and uncertain; and the
influence of the various players is unequal as
well.

b. The game is affected by anticipated
repetitive play, but the repetition is never
exactly the same and the discounting is
uncertain.

c.　Each player's decisions usually involve departmental/functional subgroups.

d.　If group incentives are present, each player has to estimate and to balance the departmental/functional vs. the group-based payoffs.

e.　There may be changes in rules of play and relevant corporate culture.

6.3　Game theory suggests a number of actions for higher management and marketing management to better understand the new decision process and environment.　These include:

a.　Making the process as much a non-zero-sum game as possible, assuring each player a positive expected payoff commensurate with expected costs, inefficiencies, and risks.

b.　Giving each player "credit", in current and future play, for "sacrifices" made.

c.　Making interdepartmental, interfunctional payoffs more uniform and certain.

d.　Stimulating cooperation via information, meetings, pressure, and the required explicitness.

e.　Encouraging the use of a marketing planning method that anticipates, addresses, and tries to preempt the concerns of each department and functional area.　(A marketing department financial specialist is often helpful).

f.　Reducing the incidence and effect of anti-marketing coalitions.

7. The Planning Model and Computer-Assisted
   Procedure

Appendix A provides a description of the

planning model/procedure and a step-by-step summary of the computer-assisted steps involved. The model content and protocols can also be visualized from the sample plan results below.

8. The Planning Procedure: Output

Table 1 provides a simulated example of the annual promotion plan for 1994, month by month, including the effect on the plan of the "nonmarketing" costs. Table 2 summarizes the dollar increases and decreases in "nonmarketing" costs resulting from the proposed marketing plan (the net result being a cost increase of $1,821,165 and a corresponding decrease in projected profit). Table 3 summarizes the per-unit increases and decreases in nonmarketing costs. This illustration is the simulated planning situation for a major brand of cereal.

9. Analysis of the Illustration in this Paper

In the cited case example, if the planning had been done without considering the cost increases for nonmarketing areas, the marketing communications plan would have been evaluated as though it would yield 9.7% more profit. However, the "nonmarketing" perspective would presumably have come into the planning eventually, one way or another.

10. Inferences for Marketing Planning

10.1 Examples of changes in marketing communications which tend to limit increases in nonmarketing costs, and thereby increase profits, while still potentially representing a good plan, include:

a. The timing and sizing of promotions so that sales/demand is less concentrated in busy times.

TABLE 1 - BRANDED CEREAL

MARKETING PLAN WITH AFFECTS ON NONMARKETING COSTS CONSIDERED

1994 - BY MONTHS

	Jan.	Feb.	Mar.	Apr.
Total Market (cases-millions-trend)	7,250	7,250	7,250	7,250
Seasonal Index	109.1	94.5	116.3	86.1
Tot. Market (cases-deseasonalized)	7,910	6,851	8,431	6,242
Brand Share (calculated)	.0194	.0120	.0289	.0190
Unit Sales (cases)	151,424	82,367	243,538	118,809
Price ($ - per case)	79.20	79.20	79.20	79.20
Cost of Goods Sold (per case,base)	37.30	37.30	37.30	37.30
Cost of Goods (direct change* per case due to promotion & advert.)	+.30	00	+2.03	00
Cost of Goods ($ case/adjusted)	37.60	37.30	39.33	37.30
Gross Margin ($ case)	41.60	41.90	39.87	41.90
Sales ($ millions)	11,992	6,523	19,288	9,409
Cost of Goods Sold ($000)	5,693	3,072	9,578	4,431
Gross Profit ($000)	6,299	3,451	9,710	4,978
Expenses ($000): Advertising ($000)	1,430	1,300	1,560	1,300
Consumer Promotion ($000)	2,356	167	1	102
Trade Promotion ($000)	73	67	2,924	585
Overhead ($000)	806	806	806	806
Total Expenses ($000)	4,665	2,340	5,291	2,793
Operating Profit ($000)	1,634	1,111	4,419	2,185
Interest Cost ($000)	480	261	772	376
Net Profit Before Taxes ($000)	1,154	850	3,647	1,809

* Estimated changes in cost of goods sold attributed to advertising and promotion.

(TABLE 1 - continued)

May	June	July	Aug.	Sept	Oct.	Nov.	Dec.	Year
7250	7,250	7,250	7,250	7,250	7,250	7,250	7,250	87,000
115.4	99.9	104.0	77.3	108.6	94.2	91.6	103.0	100.0
8367	7,243	7,540	5,604	7,874	6,830	6,641	7,467	87,000
.0180	.0147	.0199	.0148	.0314	.0205	.0199	.0120	.0195
150765	106498	150307	83,003	247,723	140,129	132,204	89,528	1696295
79.20	79.20	79.20	79.20	79.20	79.20	79.20	79.20	79.20
37.30	37.30	37.30	37.30	37.30	37.30	37.30	37.30	37.30
+.30	00	+.29	00	+2.08	00	00	00	+.67
37.60	37.30	37.50	37.30	39.38	37.30	37.30	37.30	37.97
41.60	41.90	41.61	41.90	39.82	41.90	41.90	41.90	41.23
11941	8,434	11,904	6,574	19,619	11,098	10,470	7,090	134,342
5,669	3,972	5,650	3,096	9,755	5,227	4,931	3,339	64,413
6,272	4,462	6,254	3,478	9,864	5,871	5,539	3,751	69,929
1,430	1,300	1,300	1,300	1,560	1,300	1,430	1,300	16,510
2,621	1,400	2,480	630	47	289	1,843	565	12,501
52	0	290	320	3,163	504	400	619	8,997
806	806	806	806	806	806	806	806	806
4,909	3,506	4,876	3,056	5,576	2,899	4,479	3,290	47,680
1,363	956	1,378	422	4,288	2,972	1,060	461	22,249
478	337	476	263	785	444	419	284	5,375
885	619	902	159	3,503	2,528	641	177	16,874

TABLE 2

SUMMARY OF ESTIMATED INCREASES AND DECREASES IN NONMARKETING COSTS
AS RESULT OF VOLUME, TYPE, AND CONTENT OF MARKETING COMMUNICATIONS
(1994 PLAN BY MONTH)

Cost Category

Month	Manufac-turing*	Raw Materials*	Inven-tory*	Other Logistics*	Financial*	Total
January	$ +45,055	--	$ +3,755	--	$+29,000	$+77,810
February	--	--	--	--	--	--
March	+457,469	$ -24,350	+38,113	$+24,350	+291,000	+786,582
April	--	--	--	--	--	--
May	+42,106	--	+3,509	--	+25,000	+70,615
June	--	--	--	--	--	--
July	+40,056	--	+3,338	--	+24,000	+67,394
August	--	--	--	--	--	--
Sept.	+476,090	-24,350	+39,674	+24,350	+303,000	+818,764
October	--	--	--	--	--	--
November	--	--	--	--	--	--
December	--	--	--	--	--	--
Tot.	$+1,060,776	$-48,700	$+88,389	$+48,700	$+672,000	$+1,821,165

*
 Changes are from what the "optimal" plan/proposal would be were
 "nonmarketing" costs not considered.

TABLE 3

*CHANGES IN COST OF GOODS SOLD PER CASE ATTRIBUTED TO PROMOTION

AND ADVERTISING, 1994 PLAN BY MONTH

(cost changes, plus and minus, per case**)

Cost Category

Month	Manufacturing*	Raw Materials*	Inventory*	Other Logistics*	Financial*
January	$ + .30	--	$ + .02	--	$ 29,000
February	--	--	--	--	--
March	+ 1.87	$ - .04***	+ .16	$ + .04***	291,000
April	--	--	--	--	--
May	+ .28	--	+ .02	--	25,000
June	--	--	--	--	--
July	+ .27	--	+ .02	--	24,000
August	--	--	--	--	--
September	+ 1.92	- .04***	+ .16	+ .04	303,000
October	--	--	--	--	--
November	--	--	--	--	--
December	--	--	--	--	--

*
The cost changes in this table are from what the plan/proposal would be were nonmarketing costs not considered. (The direct costs of the promotion and advertising are excluded here and included in the basic plan shown in Table 1).

**
Change in financial cost in total dollars, not per case.

***
Due to limited information, changes in costs of raw materials and "other logistics costs" are assumed to balance at 2% of cost of goods sold.

b.   Using types of promotion that limit
irregularities in sales/demand (e.g., promotions
not oriented to immediate purchases of the
product by the consumer and trade; events
marketing rather than other promotions; and
image advertising rather than direct-action
advertising).

c.   Using marketing communications' content
which lessens the adverse effect of nonmarketing
costs by stimulating: (1) purchase by mail
rather than from retailers, (2) purchase by
credit card, (3) giving less emphasis to
product/brand   variety   or   special
features/package.

d. Using marketing communications strategy and
content whose results are relatively more
predictable.

10.2   On the other hand, marketing
communications actions that actually tend to
decrease some of the nonmarketing costs include:

a.   Those that increase sales so strongly that
they stimulate lower costs and increased
efficiency.

b.   Brand-equity advertising and other longer
term communications.

c.   Most events-marketing.

d.   Publicity and public relations.

e.   Purchases for cash rather than on credit.

10.3 In the illustrated situation, a reanalysis
of some past promotional activity which now took
into account explicitly the probable cost
changes in nonmarketing activities revealed that
one prior trade promotion which was calculated
to be quite profitable, in terms of the
incremental gross margin from sales that it
brought in vs. its costs, was actually only

barely profitable. Another past promotion, which was thought to be somewhat successful in terms of incremental gross margin vs. direct costs, was now seen as a loss. Equally important, the more highly successful the promotion in bringing in incremental sales, the more its true profitability would tend to have been decreased because of the addition to nonmarketing costs.

However, in some of the promotions the addition to nonmarketing costs was not an important factor. In addition, it is known that many trade and consumer promotions are deemed successful even if they do not break even on their direct costs, because of their role in helping achieve the brand's objectives for other aspects of communications, as well as the objectives for market share, distribution, morale, etc.

11. Suggestions for Further Research

Aside from additional testing of the model/method, further study is needed to provide more complete specification and better quantification of the cost effects, on "nonmarketing" departments and functions, of marketing communications volume, type, and content. Even when this is done, special adjustments for the particular planning situation seem likely to continue to be needed.

Study is needed about the possibility that the consideration in the model/method may facilitate re-engineering of the organization's operating level to a lower and cheaper level.

In addition, free cash flow and channel-member operations will be included in planned additions to the model/method.

Study is needed to try to evolve a formal game-theoretic model of the planning and decision making process, and the equilibrium and the optimal strategies for the marketing department, for other departments and functions, and for the organization as a whole. This may provide further insights for the marketing communications planning process.

12. Conclusions

This paper describes a model and computer-assisted method for better integrating marketing communications (mainly the advertising and promotion aspects) with the "nonmarketing" aspects of the organization by including consideration of how the volume, type, and content of marketing communications increase/decrease the costs of the various "nonmarketing" functions of the organization. In the model/method, this "nonmarketing" coordination factor is added to a prior computer-assisted planning model in which the model operator simulates alternative marketing, advertising and promotion plans by varying them from time period to time period given assumptions about the market, market share, and the sales effect of different levels of advertising, promotion and price on sales. Among the important shorter term cost changes for "nonmarketing" departments and functions (mostly cost increases) are costs of manufacturing, raw materials, inventories and other logistics, financing, and servicing. Through the model and method, the specific cost effects and profit results can be estimated.

The model and method are consistent with developing trends in marketing management toward decentralization, questioning the profitability of sales promotion, cross-functional planning, profit-oriented marketing communications planning, and channel-member relationship modeling.

It is noted that nonmarketing cost considerations tend to favor certain kinds of advertising and communications strategies and tactics and to disfavor others.

13. APPENDIX A

The following is a description of the model and
the step-by-step planning procedure and
analytical process. Some of the specifics are
omitted. A sample of part of a planning result
is shown in Tables 1, 2, and 3.

The Model Concepts and Operating Procedure:
   The operator/planner accesses the program
via a personal computer diskette and is
instructed by the program/menus.
Inputs
   After an introductory message, to input the
data for the product/brand (monthly or yearly
depending on whether the annual plan by months
or the forthcoming five-year plan is being
used), the manager/operator is asked by the
program to input the information and assumptions
needed. The following inputs are needed (in
this example planning for one year, month by
month, is assumed):
1. The base manufacturing cost per unit for the
product for each of the 12 months (of the
planning period) i.e., assuming standard/ base
production volume.
2. The base cost of raw materials and supplies
per unit for each of the 12 months (i.e.,
assuming base production volume).
3. The base inventory carrying cost per unit
for each of the 12 months.
4. The base "other logistics cost" per unit for
each of the 12 months.
5. Any "other variable costs per unit" for each
of the 12 months i.e., any added category costs
due to the plan.
6. The change in cost of goods sold (either
input via management estimates or as estimated
via a subroutine which estimates it). (Please
see also Table 2).
7. The financial cost (mainly accounts
receivable less accounts payable) excluding
costs of financing inventory (but can include
the costs of financing the various promotion
elements and any other investment).

8. The overhead for each of the 12 months.
9. The size of the market in units (for all competitors) for each of the 12 months. (This is the annual market size divided into months according to the seasonal index of demand). This would reflect any growth in the market anticipated. (For analytical purposes, alternative market estimates are also used).
10. The brand/product's market share for the year is entered (can differ by month if appropriate). If no market share changes are inputted at this point (i.e., prior to the specification to follow of the tentative marketing plan for the year), either the prior year's share is used initially or the share is estimated from one of the other (annual-plan-without-nonmarketing-cost-adjustments) programs.
11. The past (prior time period's) price per unit, by month.
12. The past (prior time period's) dollar advertising expenditure by month.
13. The past (prior time period's) dollar promotion expenditure (consumer and trade) by month.
14. The program then calculates a profit and loss statement for the planning year by month, and the yearly total. This initial or base plan, in effect, indicates what is expected to occur if the past period's marketing plan is unchanged in the forthcoming market and competitive environment projected. The output content and format are similar to that in Table 1.
15. The operator can get definitions and explanations of the elements of the output. Then, the manager/operator can override any of inputs/projections, and do sensitivity analysis on the base plan, before simulating the first trial alternative marketing plan itself.
17. The next (and key) step in the planning process permits the manager/operator to vary the advertising and promotion inputs (and also vary the brand's price, if desired) to try to improve the initial ("base") plan according to whatever of the following criteria he/she may have in

mind, (i.e., net operating profit, market share, unit or dollar sales, cost reduction). If the "what if" plan involves a change in the dollar volume of the promotion or advertising in any month, the operator simply inputs this new dollar level when prompted by the program menu. If the plan involves a change in the type of marketing communications but no expenditure change, the sales effect can be adjusted by changing the elasticity response coefficients (described below) to indicate the amount and timing of the response.

18. If the operator is making/trying-out any changes in advertising, promotion, or price from the initial plan, he/she must also specify an elasticity/response coefficient (for each month) when prompted by the program. These elasticity coefficients represent the operator's experience or judgement about what the sales response to the changes in the marketing communications plan or the price will be. (In the case example which follows, the sales response to marketing communications volume changes was believed/known to be +0.46 for consumer promotion, +.39 for dealer (trade) promotion and +0.41 for promotion as a whole, weighted by the relative amount of each type of promotion. The elasticity coefficient for advertising was +.14 and -1.4 for price. Some elasticity variations by month were used, reflecting the variations in the distribution of promotion dollars between trade promotion and consumer promotion and reflecting the type of promotion (e.g., lower for consumer coupon vs. equivalent price-off; lower for very large dealer discount promotion vs. lesser-sized dealer purchase discount).

When the manager/operator inputs a change potential change in the marketing plan that will affect "nonmarketing" costs (e.g., any of the cost of goods sold elements or the fixed overhead), then he/she must presently input this separately when prompted. The cost changes in the cited example are shown in Tables 1 and 2.

A time lag for sales effect can be selected; usually zero time lag is chosen.

Bibliography

Applegate, Linda (1992), Frito-Lay, Inc. A Strategic Transition (C). Boston, Mass.: Harvard Business School.

----- (1993), Frito-Lay, Inc. A Strategic Transition (D). Boston, Mass.: Harvard Business School.

Belch, Michael & George Belch (1995), Introduction to Advertising and Promotion. Homewood, Ill.: Irwin.

Business Week Magazine (1994), "Plumper Profits Ahead: Vigorous Cost Cutting and Global Expansion Should Beef Up Results", Business Week. January 10, 1994, pp. 78-.

Cremer, J. (1986), "Cooperation in Ongoing Organizations". Quarterly Journal of Economics, CI:Issue 1, 33-49.

Eskin, Gerald (1970), Concorn Kitchens. Palo Alto, CA.: Stanford University.

Forges, Francois (1993), "Some Thoughts on Efficiency and Information", in Binmore, Ken et. al (Eds.) Frontiers of Game Theory. Cambridge, MIT Press, pp.133-148.

Fudenberg,Drew. and Jean Tirold (1991), Game Theory. Cambridge: M.I.T. Press.

Hammond, Peter J. (1993), "Aspects of Rationalizable Behavior" in Binmore, Ken et. al (Eds.) Frontiers of Game Theory. Cambridge, MIT Press, pp. 277-305.

Jaffe, Morton I. (1985), "Computer Assisted Advertising Planning: Spread Sheets and Beyond" in Stephens, Nancy (Ed.), Proceedings of the 1985 Conference of the American Academy of Advertising. American Academy of Advertising, pp. 171-174.

----- (1995), Telephone interviews with executives of various organizations.

Moulin, Herve (1988), Axioms of Cooperative Decision Making. New York: Cambridge University Press, 1988.

Myerson, Roger B (1991), Game Theory. Cambridge: Harvard U. Press, (especially chapter 10).

Radner, Roy, Roger Myerson and Eric Maskin
(1986), "An Example of a Repeated Partnership
Game with Discounting and with Uniformly
Inefficient Equilibria", Review of Economic
Studies. LIII (1) No. 172, 59-69.

Schultz, Don., et. al (1992). Integrated
Marketing Communications. Chicago: NTC Books.
Stanford University (1973), Concorn Kitchens.
Palo Alto, CA.

Wal-Mart Stores, Inc. (1994). Boston,Mass.
Harvard Business School, 1994.

Walton. Richard E. and John M. Dutton (1969).
"The Management of Interdepartmental
Conflict". Administrative Science Quarterly.
14:1:73-84.

-----, John. M. Dutton & Thomas P. Cafferty,
(1969), "Organizational Context and
Interdepartmental Conflict. Administrative
Science Quarterly. 14:4:522-541.

Modeling of Customer Response
to Marketing of Local Telephone Services

Gregory Napiorkowski and William Borghard

Consumer Lab, AT&T Bell Laboratories, USA

Abstract. The paper presents a case study of the dynamics of market response to one marketing campaign aimed at breaking the monopoly of a *local telephone service* provider by new entrants into the local market, i.e. *long distance telephone service* providers. As a rule, their target market includes households who are already their customers, albeit *long distance* only. The study illustrates an application of the *Classification and Regression Tree* algorithm to model customer response. The primary goal was to identify groups of customers most willing to dial around. The model was estimated based on almost 300,000 observations representing customers who had received the direct mail messages. The paper concludes with the evaluation of stability of the estimated tree-based response model, followed by the discussion of the *jacknife* as a way to cross-validate the model.

Keywords. direct marketing, response models, classification and regression tree, jacknife, bootstrap

1 Introduction

The paper presents a case study of the dynamics of market response to one marketing campaign aimed at breaking the monopoly of a *local telephone* service *provider*. The new entrants into this market, i.e. *long distance telephone service (LD)* providers, who are currently launching similar campaigns are taking advantage of the recent profound regulatory change in the U.S. telecommunication industry[1]. They also benefit from the fact that their brand names are already, in most cases, very well established. As a matter of fact, their target market includes households who are already their customers, albeit *Long Distance* only. The stakes are high[2] and the *Local* market is quickly becoming a new telecommunication battlefield for long distance carriers and local telephone companies. Presently, the *Local* market is open

---

[1] *Long distance telephone service* providers are now allowed to offer telecommunication services also within the administrative units called Local Access Transport Area (LATA). We realize that the telecom jargon sounds awkward, so from now on we will refer to *intraLATA* calls as *Local*; all other calls will be referred to as *Long Distance*.

[2] Most estimates place the potential revenue from residential customers within the $13 - $16 billion range.

to competition in almost all states. However, for the time being, users have to dial a number of extra digits, i.e. a chosen long distance company's access code, in order to have their calls carried by it and not by a local telephone company (a default). The new entrants' marketing message attempts to convince customers that an additional saving and a presumed higher quality of services are worth this trouble (*of dialing around*).

Customers' willingness to dial around, despite an obvious inconvenience, is an important issue for long distance carriers' managers. Firstly, if this willingness can be linked to some customers' characteristics, one could concentrate marketing efforts on the best "prospects", maximizing the expected return from this investment. Secondly, it is likely that users who bother to dial around will select the same long distance carrier at the time when all carriers are offered a chance of becoming default *Local* service providers.

A large scale market trial, being a subject of our study, was conducted in 1994 in one of the states. The purpose of this trial was precisely to answer the above posted question: what is customers' propensity to dial around while making *Local* calls? Our primary objective was to identify groups of customers most responsive to the advertising, and first of all, *Direct Mail* campaign urging them to dial around as opposed to customers not willing to do so.

The paper is organized as follows. Section 2 gives an overview of the experiment, focusing on timing of mass media advertising and direct marketing campaigns. Section 3 contains an aggregate analysis of their respective impacts on *Local* dial around usage patterns, such as minutes of use and call distributions by day of the week, time of day etc. The customers' consistency in dialing around while making *Local* calls is also examined here. Section 4 defines the potential explanatory variables to be used in the response models. Section 5 reviews briefly the applied methodology, i.e. the *Classification and Regression Tree* (*CART*) algorithm. Section 6 presents the *CART* estimation results and their interpretation. We include here a comparative analysis of customers' propensity to dial around in subsequent periods after having received the direct marketing message. We also address the issue of response rates among customers who did not receive direct mail (*DM*) messages and were exposed to mass media advertising only. Section 7 follows with the validation analysis and statistical evaluation of the *CART* estimates.

All numerical estimates of customers' response rates are masked. The actual rates may not be revealed in this publicly available paper.

2 The Marketing Campaign Overview

The market trial was conducted with an objective to provide the first empirical data on customer willingness to use *LD provider*'s network to make *Local* calls. While choosing the area of the trial the following criteria were considered: relatively high volume of *Local* usage, friendly regulatory environment and logistics ensuring timely collection of the trial data.

The trial marketing effort attempted to achieve three objectives:

- To make customers aware of the fact that *LD provider* offers not only *Long Distance* but also *Local* services.
- To educate customers how to access *LD provider's* network to make *Local* call.
- Convince customers that dialing additional five digits is worth the trouble.

The campaign started with TV (May 13 - June 10) and newspaper (May 16 - May 27) ads on weekdays followed by *DM* messages sent on May 23rd and May 24th to the selected group of more than 270,000 AT&T customers. From now on we will refer to this group as the *Sample*. The mass media messages were reinforced by radio ads (June 13 - June 24). The letters to customers were tailored to their calling patterns[3].The customers' response data consist of detailed records of all *Local* dial around calls made between April 18th and June 30th inclusive. The term *detailed record* indicates that we know the originating and terminating telephone numbers, the exact time calls were made and their length.

A customer is considered a *respondent* if he or she made at least one *Local* dial around call during the above period. By merging information contained in the *Sample* with the response data one can identify the following customer sub-populations:

- members of the *Sample* and responders
- members of the *Sample* and non-responders
- non-members of the *Sample* and responders

For completeness one should also mention the fourth, least interesting sub-population, namely non-members of the *Sample* and non-responders.

3 Response Data Analysis

Although our primary objective in this study was to classify customers according to their propensity to dial around to make *Local* calls, we were also interested in the aggregate customers' response to mass media advertising and *DM* messages. In particular we addressed the following questions.

---

[3]For instance, the message sent to customers enrolled in a given discount plan emphasized the fact that the *Local* portion of the bill, if added to its *Long Distance* part would allow to reach a bill size threshold, that would qualify a customer for a discount of up to 20% of the entire bill.

1. What was the volume of *Local* dial around calls immediately before, during and in the aftermath of the marketing campaign?

2. Can the impact of media advertising and *DM* be easily detected and, if yes, how long did it take customers to respond?

3. What is the dial around *Local* traffic pattern like (e.g. distribution of calls by rate periods) and how does it compare to the Long Distance traffic?

4. Does dialing around quickly become a habit or is it rather a one time adventure?

Fig. 1 and Fig. 2 contain most of the information needed to answer the above questions. The only difference between the two figures is the measure of customers' response used: number of customers dialing around at least once (Fig. 1) vs. total number of calls (Fig. 2). For clarity, not all but only some selected days of May and June are marked on the horizontal axis.

A closer look at both figures allows us to draw the following conclusions.

1. Both the number of responders and the volume of usage increased significantly at the time the TV advertising started. However, not surprisingly, the effect of *DM* was much stronger.

2. The effects of both TV and *DM* messages were virtually immediate (if one allows up to two days for mail delivery).

3. The radio commercials seemed to have little, if any, marginal impact on *Local* dial around calling in the period following the peak usage, i.e. after June 3rd.

4. The daily number of *Local* dial around callers (Fig. 1) saturated quickly, i.e. within ten days after *DM* messages had been sent. Fig. 3 helps explain how this steady state of the number of customers was reached. The number of new customers (i.e. those who use the service for the first time) stabilized quickly after the first week of June. However, these inflows of customers did not result in a systematic increase in the daily number of responders. Apparently, they were offset by a similar number of customers who, on a given day, did not make *Local* dial around calls. The pattern shown on Fig. 1 is mimicked by daily numbers of calls (Fig. 2). It implies that the average usage per customer is stable over time and, most likely, the customer distribution by usage does not change either. Searching for heavier *Local* users we found that, among all June' 94 callers, approximately 30% used the service on just one day, but the remaining 70% accounted for about 97% of recorded calls.

5. The daily usage reveals an unexpected daily pattern: the traffic on weekends is visibly lighter than on respective weekdays.

The last observation is in a sharp contrasts to the usually observed patterns, according to which a very significant proportion of calls (both local and long distance) occurred on weekends. We hypothesize that this unexpected *Local* dial around call distribution by time of day might be, among other factors, a result of:

- Mass media advertising was run on weekdays only.

- A large proportion of *responders* had business related calling needs (errands, work at home, etc.) that had to be taken care of during the day.

- Price perception among newly recruited *Local* customers, that local calls are much less expensive than *Long Distance* ones; It manifests itself in shifting of calls from the less convenient night/weekend period into day or evening hours. In contrast to the long distance calls, these hours are now considered more affordable.

4 Potential Explanatory Variables of Customers' Response.

The variables considered by us can be classified into four groups. We briefly describe them in the subsections below.

4.1 Long Distance Telephone Usage

We focused on the following characteristics:

- Discount plan subscription

- Telephone bill size

- Calling card usage

- Operator assisted usage

- Usage distribution by day, evening, night/weekend

4.2 LD Provider's Brand Loyalty Variables

We identified three variables as a manifestation of customers' loyalty. They represent purchases of the *LD* provider's brand products. We are not free to disclose them in a publicly available document and we will refer to them just as *Loyal_1*, *Loyal_2* and *Loyal_3*.

4.3 Demographic Variables

This group contains various demographic and income characteristics of the customers' households that were available to us. The most promising variables seemed to be:

- Occupation (head of the household's, spouse's)

- Age (head of the household's, spouse's)

- Income

- Responsiveness to marketing messages in the past

- Length of stay at the present address

4.4 Variables Describing Customer's Location

The customer's location is defined generally in terms of local population density with regard to *Local* calling area. Specifically, we tried to quantify customers' opportunities/necessities to make *Local* calls. We believe that these characteristics are major determinants of the volume of *Local* calling. We constructed two variables that we found very useful in this respect. We will refer to them as *Loc_1* and *Loc_2*. In this paper we may only provide their interpretation and not exact definitions.

Loc_1 measures "local calling opportunity". This construct variable is a function of the number of telephone lines in the area, distance from customer's residence to other densely populated localities and number of area's business establishments.

Loc_2 is another composite variable. It is designed to reflect mostly the residents' phone calling patterns, as well as external estimates of the overall local market potential.

4.5 Customer's Calling Circle Variables

These variables characterize a calling circle of which a given customer is a member. Evidence abounds that members of calling circles affect each other's behavior as telecommunication users. We defined two variables to catch the calling circle effects: Community of Interest (*COI*) and Word of Mouth (*WOM*). We describe them below.

4.5.1 COI

Long distance carriers' data systems contain rich information on customers' *Long Distance* usage, including each call's detail, such as time, terminating number etc. Exploring this information, we attempted to construct, for a given customer X, a variable that, by tracking X's calling circle, links the entire circle's *Long Distance* calling pattern (i.e. "who calls whom?") to the X's *Local* usage. In general, we hypothesize that the higher the concentration of calling circles' members in a given local area, the more likely they are to make *Local* calls.

4.5.2 WOM

In this context, the word-of-mouth phenomenon is the spreading of information by customers themselves regarding the benefits of using *LD* provider's network for *Local* calling. The only word-of-mouth channel we can observe is a phone conversation itself. Hence we defined the *WOM* variable as the total number of *Local* dial around minutes a customer receives. We realize that this variable is "dynamically" defined, i.e. its magnitude increases over time. We address this issue in Section 6.4.

4.6 Pre-selected Explanatory Variables of Customers' Response

We conducted a preliminary analysis of how significant the above variables are in determining the probability of customers' response to the *DM* messages. We based our judgment on comparison of each variable's distributions for two sub-populations of customers: those who used *LD* provider's *Local* dial around service and those who didn't make any such calls. It turned out that for most "potential" variables the differences in these distributions were statistically insignificant and we eliminated them from further considerations. The list of variables that successfully passed this test is presented below.

- Discount plan subscription (*DPS*)

- Telephone bill size (*TB*)

- Calling card usage (*CC*)

- *Loyal_1*

- *Loyal_2*

- *Loc_1*

- *Loc_2*

- Word of Mouth (*WOM*)

5 Methodology

Recall here that our research objective was to classify customers in terms of their responsiveness (by making *Local* dial around calls) to *DM* messages. It was tantamount to estimating a probability of response for every customer. More precisely, our task was to specify a binary discrete choice model, because our definition of the customers' response qualified them to only two possible categories: *responders* vs. *non-responders*. The most widely applied binary discrete choice models are *logit* and *probit*. Their popularity is typically attributed to the well developed underlying probabilistic theory and, consequently, the rich statistical inference offered.

In this study we decided to use an alternative modeling technique, based on the classification and regression tree (*CART*) algorithm. The resulting models are known as *tree-based* models because they are usually presented in the form of a tree with such elements as *root*, *branches* and *leaves*. A tree is "grown" by binary recursive partitioning of a given population (*Sample*, in our case). The first partitioning is applied to the entire *Sample* (i.e. root), and creates the first two branches. The next split of these two branches follows. This process, if continued, splits our *Sample* into increasingly more homogeneous subsets (i.e. leaves) of customers in terms of their probability to respond to the *DM* messages. The algorithm assures that the most important variables determine the top branches of the tree. The leaves are defined by values of all variables that were found significant in identifying ever more homogeneous groups of customers while growing the tree. A detailed description of tree-based models and of the *CART* algorithm can be found in Breiman et al (1984) and Chambers and Hastie (1991).

Among the features of tree-based models the following two were the most appealing to us:

- Ability to capture effects of interactions of two or more variables (non-additive behavior) on probability of response

- Easiness to perform an exploratory analysis to uncover structure in data, i.e. interpretability of the results.

The first feature allowed us to greatly reduce the number of different forms of interactions between predictive variables[4] we would have to consider if we worked with the classical probability models like *logit* or *probit*.

---

[4].In this study, the predictive variables are chosen from the pre-selected explanatory variables listed above.

6 Modeling Results

We specified two versions of tree-based models, depending on whether *WOM* was available for the analysis or not. Let us note that at the time when *Local* services are just introduced to the market, there are virtually no historical data on *Local* calling and, consequently, one cannot determine the *WOM* value at the customer level. This information becomes gradually available with time. Our experience indicates that a relatively short period of four weeks can provide a substantial dose of data on *Local* calling. Because of that, the problem of identifying likely responders to *DM* messages lends itself to a two phase approach. Phase 1 will target high probability responders without taking advantage of the *WOM* variable. Phase 2 will introduce into analysis also this particular variable. By doing so one should increase the precision with which we target the most likely responders. It should be pointed out here that *WOM* is, as opposed to all other variables included in this analysis, a dynamically defined variable - its value changes with the time elapsed. It would be interesting to find out the length of the period after which the marginal impact of the updated *WOM* variable on the tree structure becomes negligible. We will address this issue in Section 6.4.

The tree-based classification results are presented on the attached Figures 4-10. For clarity, we limited the trees' sizes to twenty terminal leaves (rectangular nodes). The numbers associated with every node (oval or rectangular) are to be interpreted as follows:

- The denominator of the fraction denotes total number of customers "assigned" to a given node.

- The numerator of the fraction denotes the number of responders.

- The fraction's value is shown, in %, within the node's symbol.

The lines connecting the nodes are labeled with the variables' names and their respective values, along which the split occurred. Most variables can assume only two values: Yes(Y) / No(N) or High(H) / Low(L). Three variables: *DPS, TB* and *WOM* are not binary ones. They can respectively assume three, five and six values (categories). Values of *DPS* indicate the discount program customers can subscribe to: *Plan_1, Plan_2, Plan_3.* For *TB* and *WOM*, the higher the value (category), the larger the telephone bill or the level of word-of-mouth, respectively.

We built our tree models based on the sample of 273,777 customers who had received a *DM* message. An overall response rate within our working sample was 10.3%, which might be considered quite high for a direct marketing campaign. We should remind the reader at this point that it sufficed to make just one dial around *Local* call to be considered a responder. Unless stated otherwise, this study "offers" customers over five weeks (from May 23 to June 30) to respond. Let us now, tracing the steps of how the trees were grown, identify the most significant variables driving the customers' response.

6.1 Phase 1 Model: WOM Variable Not Available

Fig. 4 displays the tree-based response model built under assumption that no information on received *Local* calls is available. It turns out that the primary characteristic related to the higher response rate is customer's enrollment in the Plan_3 program. For the *Plan_3* customers the observed response rate (20.3%) was almost twice as high as the average (10.3%). It is not really surprising, considering that the *Plan_3* enrollees have a clear financial incentive to use *LD provider's* network also for *Local* calling.[5] We can conclude further that the second important driver of customer's response is his/her residence's location, as described by the *Loc_2* variable. The higher *Loc_2* value is associated with the higher response rate. Let us note that this variable is equally important for both participants and non-participants in the *Plan_3* program. The third and fourth characteristics that enter the picture are two AT&T brand loyalty variables. At the lower levels of the tree, a finer split between more and less likely responders can be produced by using such customers' characteristics as *Telephone Bill, Loc_1* variable as well as the fact that customers subscribe to *Plan_2* program. The relatively low rank of the latter came as a surprise to us because a participation in a *discount plan* increases, as a rule, customers' awareness of new services offered by "their" telephone companies.

The splits of nodes by respective variables are consistent with our expectations. The tree structure confirms the notion that the customers loyal to the *LD* provider's brand are more likely to respond to its messages. Also, customers with higher *Loc_1* and *Loc_2* scores are more likely to respond than customers with the lower scores. The only exception seems to be the *TB* variable. The most surprising result is that the highest (4,5) and the lowest (1) categories of this variable are sometimes grouped together as opposed to the group of middle bill size levels (2,3). The higher response rates are typically associated with the latter. One possible interpretation is that the customers with the extremely low or high phone bills are equally disinterested, albeit for different reasons, in following the phone companies' numerous marketing campaigns.

At the very minimum one would like to identify groups of customers with the higher than average probability of response. According to Fig. 4, all *Plan_3* enrollees belong here. Among the remaining customers one can identify just three clusters meeting this condition. The corresponding response rates are 15.3%, 12.5% and 11.6%. The fourth node has an estimated probability (10.7%) virtually indistinguishable from the average. In each of these cases the decisive factor was the loyalty towards the *LD* provider's brand. The dominant role in identifying the most

---

[5] Plan_3 allows customers to combine Long Distance- and Local usage to take advantage of the discounts that are based on the volume of customer generated phone traffic.

likely responders played by the enrollment in *discount plans* suggests that, in order to identify the high probability responders among the remaining customers, one should focus analysis just on them and remove the *DPS* variable from the data set. The resulting classification, although very interesting for the marketing managers, does not add much to our illustration of the tree-based classification and we will not discuss it in this paper.

6.2 Phase 2 Model: WOM Variable Is Available

The only difference between the tree-based classification results presented here compared to the previous section is that one more variable, *WOM*, was included within the set of explanatory variables. Its impact on the tree structure is shown on Fig. 5. The *WOM* variable turns out to be the second most important characteristic by which one can identify the high probability responders (the first is still enrollment in the *Plan_3* plan). Its power is reflected by the fact that it allows us to identify the numerous clusters of customers with the above average response rate (exceeding 30%), even if they are not enrolled in the *Plan_3* program. In an extreme case, the observed response rate among *Plan_3* customers who received a very high dose of word-of-mouth exceeded the 50% level. The remaining explanatory variables now play the less prominent role. For instance, *TB* now enters only at the very bottom of the display of top twenty terminal nodes.

6.3 Incremental Effects of Direct Marketing on Response Rate

One can gain an additional insight into the *DM* messages' appeal to customers by comparing the classification results for the *Sample* vis a vis customers who did not receive any *DM* message. Those (*Non-DM*) customers were only exposed to the mass media advertising and, possibly, word-of-mouth contacts. The tree-based classification models for them are displayed on Fig. 6 (phase 1 model) and Fig. 7 (phase 2 model). Note, that the percentages within nodes are the estimates of the response rates in relation to the entire population of *Non-DM* customers. They are derived by rescaling the rates observed within the Non-DM sample (fractions under nodes) that was used to specify the trees. Thus the rescaling factor is the same as the sampling rate used for *Non-DM* customers and equals 0.15.

The first obvious conclusion is that the *DM* messages increased the response rate by a factor of seven (from approximately 1.4% to 10.3%). Secondly, the primary characteristic that identifies more likely users of AT&T dial around *Local* service is *Telephone Bill*. It is not clear at this point whether this fact reflects a stronger resonance of mass media advertising among customers who use other AT&T services (*Long Distance* and/or *International*) or whether the *Local* usage is somehow correlated with other usage. Comparing further Fig. 6 and Fig. 4 one can conclude that the remaining explanatory variables have, in both cases, very similar impact on the tree-based classifications. The most significant variables are again: *DPS*, *Loc_2*, *Loyal_1*, *CC*, *Loc_1*. For instance, customers identified as loyal to the *LD* provider's brand are, as a rule, at least twice as likely to use *Local* dial around service than an average *Non-DM* customer. The only visible difference between Fig. 6 and Fig. 4 is that the Loyal_2 variable is disregarded in case of *Non-DM* customers (Fig. 6). It is probably overshadowed by a more important indicator of loyalty towards *LD* provider's brand (*Loyal_1*). For the *Sample* customers, *Loyal_2* seems to act as a reinforcement of loyalty effects already introduced by *Loyal_1* (see Fig. 4).

Yet, an overall impression is that the *Local* service offered by *LD* provider appeals to the same categories of customers among the *Sample* and *Non-DM* sub-populations. One can also conclude that the *DM* messages were very effective, if they caused an increase in the response rate for the corresponding customers' clusters by, on average, a factor of 7 and, in a few instances, even higher.

A closer review of the left branch of Fig. 6 makes us suspect that some customers were wrongly labeled as zero users. Otherwise it is hard to explain why so many zero users (1382 in our case) were interested in the *Plan_3* discount program.

A tree-based classification displayed on Fig. 7 confirms unequivocally the importance of the WOM variable in identifying the most likely *Local* responders. Not only the primary split separates customers into low and high volume receivers of *word-of-mouth*, but the finer split of the latter is, for two more levels of the tree, also based on this customers' characteristic! Only then is there room for contribution of other variables, like *TB*. One should emphasize that the observed response rate among the top group of receivers of *word-of-mouth* contacts is 7.7%, i.e. more than five times the average.

The left hand side of the tree on Fig. 7 displays a classification of customers who received a minimal, if any, number of *word-of-mouth* contacts. This part of the tree is, not surprisingly, very similar to Fig. 6.

6.4 The Classification Dynamics

A tree based algorithm offers a static picture of customers' classification with respect to their response rate to the *DM* messages. This static picture corresponds to a given period of time. In this study it is over five weeks long (4/24/94 - 6/30/94). Hence all estimates of response rates for different clusters of customers shown so far are to be interpreted in terms of five weeks. The stability of our classification for a given period will be discussed in Section 7.

Here we will focus on another related issue: how the length of the chosen response period affects the results of our analysis. More specifically, we would like to know the answer to the following questions:

• How does the response rate evolve over time?

- Were the changes in response rates uniform across the major identified customers' clusters?

- Is the tree-based classification stable over time?

- Does the effect of WOM, the only "dynamically" defined variable, diminish with time?

In order to answer the above questions, we developed tree-based classifications for three sub-periods: May 23 - June 6, June 7 - June 20 and June 21 - June 30. We decided to split the response period into natural calendar units of two weeks each. Therefore the third sub-period contains, by default, only ten days.

The respective tree models are displayed on Figures 8, 9 and 10. The answer to the first above question is straightforward: the vast majority of responders (about 2/3 of them) became ones during the first two weeks after having received the *DM* message. Then the response rate drops rather rapidly, to barely reach, in the third sub-period, just 1/7 of the response rate observed for the first sub-period.[6]

The decline in the response rate over time is quite uniform across different customers' clusters, although *Plan_3* enrollees tend to respond marginally faster than the rest of the *Sample*.

The question about the models' stability is equivalent to asking whether the relative importance of explanatory variables, vis a vis each other in the models' tree structure, remains the same as time elapses. In general this is the case, but certain nuances can be observed.

Comparing the first sub-period (Fig. 8) and the overall results (Fig. 5) leads to the conclusion that, although both tree classifications are strikingly similar, a different loyalty variable helps identify the most likely early responders.

After two weeks (see Fig. 9), the *WOM* variable loses its appeal slightly among *Plan_3* customers and we should rely more on the customers' location characteristic. Among *Non-Plan_3* subscribers one should focus first of all on receivers of *word-of-mouth* contacts, and then on the value of loyalty variable - the location characteristic becomes less important.

Finally, after four weeks from the *DM* mailing, the most promising variable, next to *WOM*, becomes *TB*. In general, the higher response rates are observed for customers with the broadly defined medium telephone bills, although sometimes the *CART* algorithm splits the nodes along unexpected *TB* lines (e.g. putting together categories

---

[6] This is a conservative comparison, as we re-scaled the third sub-period rate (0.9%) to account for a number of days. By doing so we are essentially stating that it is about 1.2%.

2 and 5). We believe this is due to inaccurate records of phone bills and/or to the fact that heavy international users have, as a rule, high telephone bills, but they pay little attention and do not respond to the advertising of domestic services.

One can also conclude that the importance of the *WOM* variable does not fade with time. With the exception of *Plan_3* customers in the second sub-period, the volume of received "word-of-mouth" contacts is consistently the best identifier of high probability responders.

We found it surprising that, contrary to our expectations, calling card users are typically no more likely to make a dial around *Local* call than customers who do not use calling card services. We had hypothesized that the former, being exposed to a necessity of dialing additional (and numerous) digits would be less averse to dial the *long distance carrier's* access prefix as well.

7 Response Model Validation

Response models are usually validated by so called "lift factors" (also known as gains chart factors). When targeting specific customers, the lift factors show by how much one can increase the expected response compared to the case when *DM* messages are sent out randomly. The lift factors are typically calculated for deciles of the customer population ranked according to decreasing probability of response. Naturally, the top deciles' lift factors provide the most interesting information.

Yet, most classical classification models are evaluated in terms of two classification errors: *false positive* and *false negative*. *False positive* error shows what fraction of customers declared (based on adopted rule) as responders fail to do so. *False negative* error reflects probability of an opposite undesired event that customer did respond to the message, although, according to our model's rule, he/she should not have done it. Despite the fact that we consider these classification errors less informative than lift factors as response model's "validators", we follow the tradition and enclose their levels, next to the lift factors' estimates, in the Section 7.1 below.

7.1 Resubstitution Estimates of Lift Factors and Classification Errors

In this study we attempted to identify all, even if not heavily populated, clusters of the most likely responders to the *Local* direct marketing messages. Having set forth this objective, we decided to use all available information on customers-responders. So at first we evaluated the model's performance based on the sample used to build it.

In case of the tree classification model displayed on Fig. 5, the cumulative lift factors for the first, second and third deciles are: 2.32, 2.03 and 1.77. It means that one can increase the expected response rate, compared to the 10.3% average, by a factor of 2.32 (i.e. up to 24%), if just top 10% of the customer base are targeted. The second lift factor (2.03) has the same interpretation, but is applied to the top 20% of the customer base etc. Based on these levels of lift factors we consider the derived customers' classification as satisfactory.

The *false positive* and *false negative* errors for the same model can be derived from TABLE 1.

TABLE 1

Actual Response	Probability of Response		
	< 30%	>= 30%	Total
Yes	237857	3064	240921
No	30635	2221	32856
Total	268492	5285	273777

If we use the rule that the customers for whom the estimated probability of response exceeds 30% are the responders, then the *false positive* and *false negative* errors are 58.0% (3064/5285) and 11.4% (30635/268492) respectively. One could obviously adopt another rule resulting in different levels of these errors (there is a trade off between them). Note that relatively low *false negative* error still leaves us with a substantial number of customers who are responders, but are not recognized as such.

Although the resubstitution estimates are commonly used in the model validation they should be considered as measures of last resort. Their obvious deficiency is that they present too optimistic picture of a given model's capabilities. For this reason we extended our analysis by an approach known as *V-fold cross validation* (see Breiman et al (1984)).

7.2 V-fold Cross-Validation of Lift Factors - An Outline

The outline of the *V-fold cross validation* presented here follows Section 1.4 in Breiman et al (1984). The focus on *lift factors* is a consequence of our earlier stated belief that they are more useful than *classification errors* in the evaluation of the response models.

The observations in the *Sample* are randomly divided into V subsets of equal sizes. One subset (say, j) is held out and the tree classifier is grown based on the remaining $V - 1$ subsets. Then the *lift factors* are calculated based on the j-th subset. This procedure is repeated for each subset, resulting in V estimates of *lift factors* for every decile. An *out of the sample* estimate of lift factor LF_I is an average of LF_{ji} over $j=1,2,...,V$, where $I = 1,2, ..., 10$ stands for an i-th decile.

The underlying assumption of cross-validation is that the classifiers are "stable". It means that the estimated lift factors should be symmetrically distributed around their means with the "reasonable" variation. It seems that the stability requirement is fulfilled if the trees grown for each of the *V* steps are similar in terms of how their *branches* and *nodes* close to the *root* are defined.

What remains to be decided is the value of *V*. Breiman et al (1984) experimented with *tenfold cross-validation* to estimate the tree misclassification rate and concluded that "the resulting estimators have been satisfactorily close to $R(d)^7$ on simulated data".

7.3 V-fold Cross-Validation of Lift Factors - Results

One can observe that the *V-fold cross-validation* procedure is conceptually similar to the *delete-d jacknife*. The latter can be thought of as a generalized jacknife: instead of dropping just one observation at a time from a given sample one leaves out *d* observations. Note that the *delete-d jacknife* does not require that the *d*-element left out subsets be mutually exclusive. Altogether, one can generate C_d^n such subsets (*n* is a number of observations in a sample). When *n* is large, so is the number of subsets. In such a case the recommended approach is to work with the randomly selected manageable number of these subsets. The *jacknife* principle remains the same, i.e. the estimator or, in our case, a tree growing algorithm is always applied to the observations remaining in the sample.[8]

The evaluation of our tree based response model should provide an answer to two basic questions:

• What are the "true", i.e. estimated based on the left out observations, levels of the *lift factors*?

• What are the standard errors of the *lift factor* estimates?

To answer these questions we experimented with the *V-fold cross-validation* and *delete-d jacknife* algorithms. We conducted our analysis for the following three values of *V*: 2, 4 and 10.

Let us focus on the *V* = 4 case as an example. The *Sample* was randomly split into four mutually exclusive subsets with the same number of observations (68444 = 273777 / 4). Three of these subsets together were used as a learning sample to grow the tree. The remaining fourth subset was used to calculate the *lift factors*. This process is to be repeated four times for one random split of the *Sample* and it represents one *iteration*. Each iteration produces four estimates of the *lift factors* for

---

[7]$R(d)$ is their notation for the "true" misclassification rate.
[8]For further details see Section 11.7 in Efron and Tibshirani (1993).

every decile. We conducted thirty such iterations. It follows that for each decile we had 120 estimates of the respective *lift factor*.

For $V = 10$, we have ten estimates of every *lift factor* per iteration. Finally, there are only two such estimates per iteration if $V = 2$. Also for each of these cases we conducted thirty iterations.

For every V and for every decile we calculated, based on all available estimates, the average estimate of the respective *lift factor* as well as its standard error. As far as the latter is concerned, we followed the spirit of the jacknife algorithm. It means that we first calculated standard deviations for every iteration and then found their average[9]. The results are presented in the TABLE 2 and TABLE 3 below. To offer additional insight, Fig. 11 shows, as an example, the empirical distributions of the top decile *lift factor* for three values of V. This picture is representative for other deciles as well. One can make the following observations based on TABLES 2 and 3:

- The *lift factors'* averages do not depend on value of V

- The higher the V, the higher are the standard errors of the *lift factors'* estimates.

TABLE 2

LIFT FACTORS' AVERAGES

Decile	$V = 2$	$V = 4$	$V = 10$
1	2.1636380	2.1699019	2.1685233
2	1.6889115	1.6817965	1.6851182
3	1.2098822	1.2162576	1.2198065
4	0.8478134	0.8463273	0.8404986
5	0.8029222	0.7986622	0.8009320
6	0.7791341	0.7768549	.7706774
7	0.6272074	0.6287616	0.6263751
8	0.6277343	0.6270782	0.6274157
9	0.6265236	0.6291275	0.6311388
10	0.6261972	0.6251275	0.6293091

---

[9] We also calculated standard deviations for each *lift factor* based on its estimates across all iterations. The numerical results are very similar.

TABLE 3

JACK KNIFE ESTIMATES OF LIFT FACTORS' STANDARD ERRORS

Decile	V = 2	V = 4	V = 10
1	0.02106940	.03625474	0.06732060
2	0.02227592	0.03641882	0.05818633
3	0.02914356	0.03304510	0.05446012
4	0.02042446	0.03378421	0.05087058
5	0.01924067	0.02613168	0.04281687
6	0.02771656	0.02771656	0.04506758
7	0.01609589	0.03064207	0.04193554
8	0.01463244	0.02809452	0.04591909
9	0.01604115	0.02910146	0.04267886
10	0.01647113	0.02555555	0.04209083

The second observation requires some further discussion. Note that the *lift factors'* volatility is a result of a superposition of two volatilities: within the subsample used to grow the tree and within the remaining subsample used to estimate the *lift factors*, i.e. a test sample. Apparently, at least for the *V* cases considered by us, the increase in the size of test sample significantly lowers the estimated standard errors. On the other hand one has to remember that the *jackknife* estimators of standard errors understate their "true" values. Consequently, appropriate "inflation factors" have to be applied to these estimators. It is not clear to us what inflation factors should be used for standard errors shown in TABLE 3. The *bootstrap method* of cross-validation could probably offer some suggestions in this respect. However, it is very computer intensive and time consuming and, at this point, such an effort would go beyond assumed scope of this analysis.

Yet, one can conclude that even conservative estimates of *lift factors'* standard errors are relatively not high and the presented response model is a reliable tool of targeting the most promising customers.

8 Conclusion

The paper discusses the customers' willingness to dial a prefix code (i.e. dial around) in order to place *Local* calls on *long distance* rather than *local telephone service provider's* network (a default). The study is based on data from the *Local* market trial, conducted in 1994 in one of the states. The main objective was to identify, using available customers' characteristics, the groups of customers who are most likely to respond to the direct marketing messages urging them to dial around while making *Local* calls.

We found that the following aspects were the most appealing to customers and helped them "put up" with the inconvenience of dialing around:

- Financial incentive (enrollment in the discount plan, like *Plan_3*)

- Location of customer's residence (it imposes a necessity to make *Local* calls)

- Loyalty to the *LD provider's* brand

- Persuasion by other users (*word-of-mouth*)

We found the *word-of-mouth* to be the most influential variable determining the customer's probability to respond to the direct marketing messages.

There is no question that, to better understand the customers' behavior in the *Local* market, further studies are required. Once more information becomes available, such issues like the customers' responsiveness to the follow up (second "wave") advertising, the extent of response (i.e. all or just some *Local* calls are dialed around) and across states differences/similarities should be addressed.

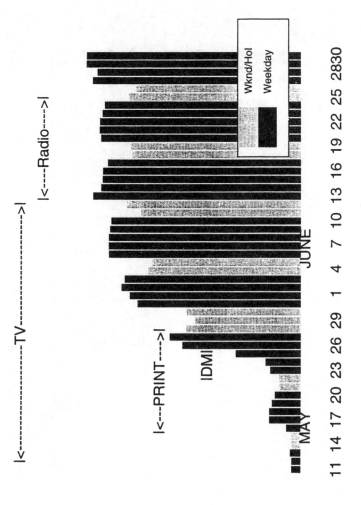

- Fig. 1 -

272

DIAL-AROUND LOCAL CALLS

Dial-around calls

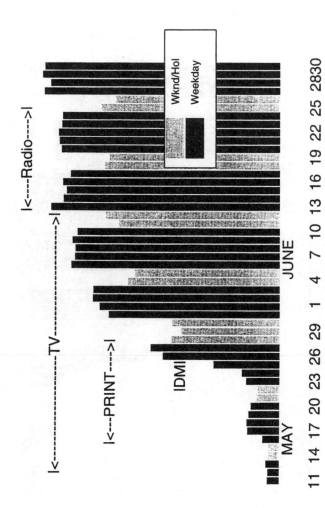

- Fig. 2 -

273

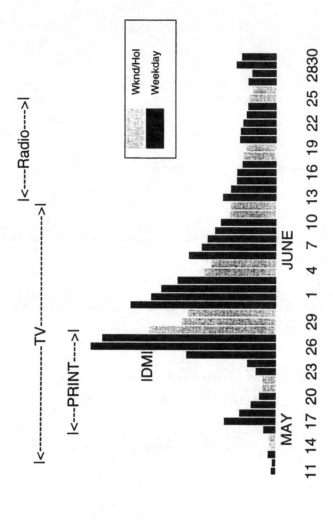

- Fig. 3 -

274

Response prior to 7/1/94: Phase 1

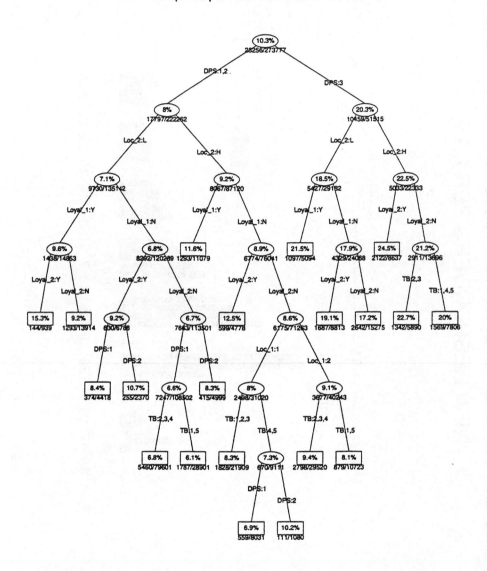

- Fig. 4 -

Response prior to 7/1/94: Phase 2

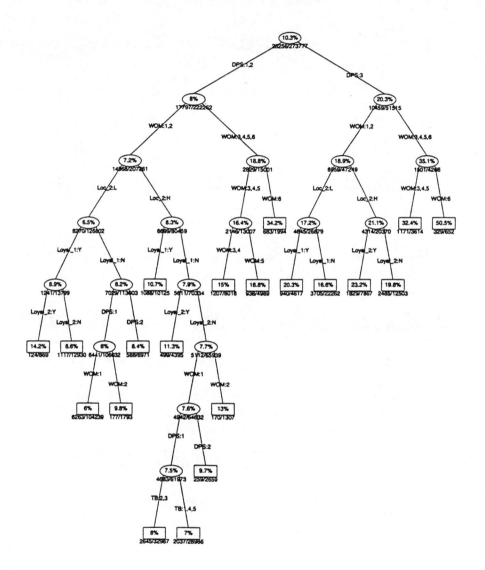

- Fig. 5 -

Response prior to 7/1/94: Non-DM, Phase 1

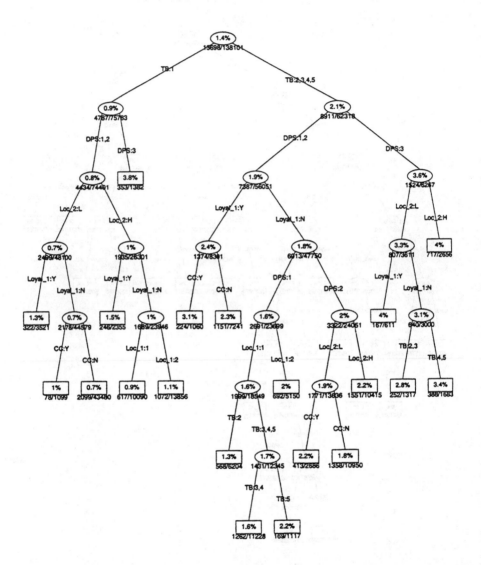

- Fig. 6 -

277

Response prior to 7/1/94: Non-DM, Phase 2

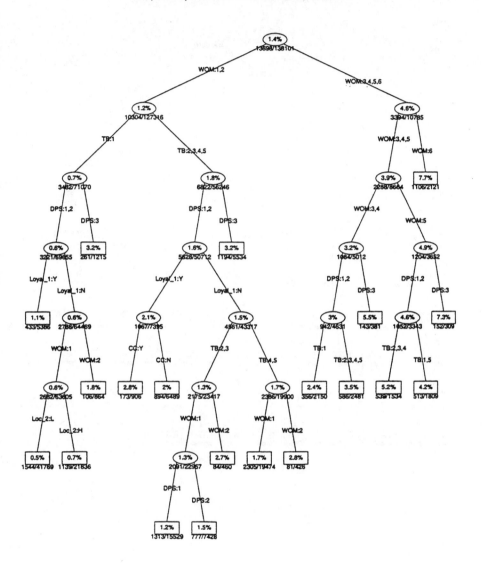

- Fig. 7 -

Response prior to 6/7/94: Phase 2

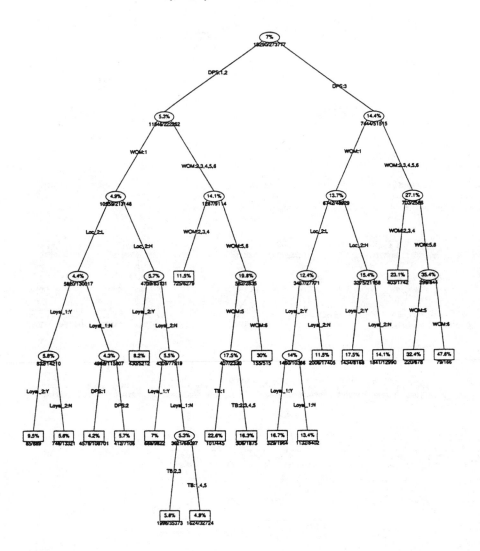

- Fig. 8 -

279

Response 6/7/94 - 6/20/94: Phase 2

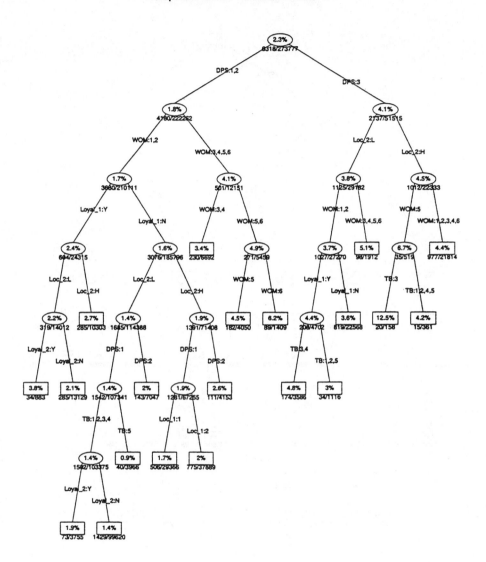

- Fig. 9 -

280

Response 6/21/94 - 6/30/94: Phase 2

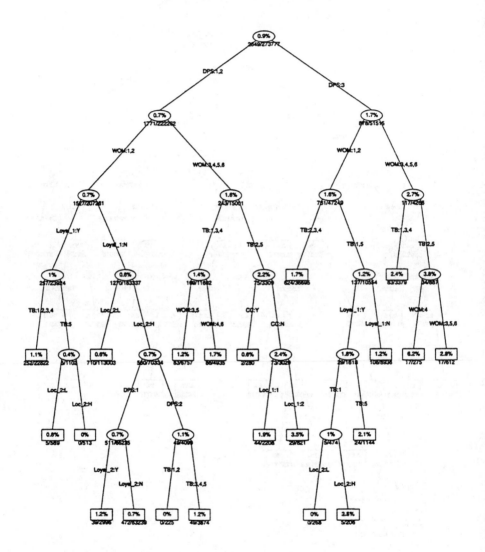

- Fig. 10 -

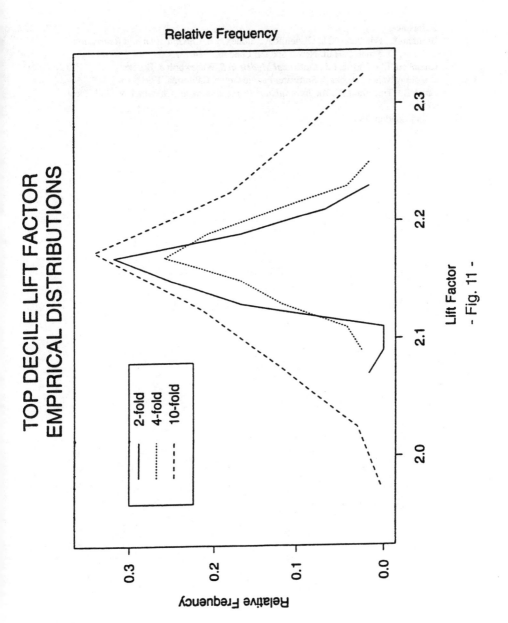

TOP DECILE LIFT FACTOR
EMPIRICAL DISTRIBUTIONS

Lift Factor
- Fig. 11 -

References

Breiman,L., Friedman,J.H., Olshen,R.A., Stone,Ch.J., *Classification and Regression Trees*, Chapman & Hall, New York and London, 1984

Chambers,J.M, Hastie,T.J., *Statistical Models in S*, Wadsworth & Brooks/ Cole Advanced Books & Software, Pacific Grove, California, 1991

Efron,B., Tibshirani,R., *An Introduction to the Bootstrap*, Chapman & Hall, New York
and London, 1993

LIST OF CONTRIBUTING AUTHORS

Bandyopadhyay, Subir
Faculty of Management, McGill University,
1001, Sherbrooke West,
Montréal, H3A 1G5   Canada

Borghard, William
AT&T Bell Laboratories,
600 Mountain Avenue,
Murray Hill, NJ 07974-0636   USA

Breton, Michèle
GERAD, HEC
5255 Decelles,
Montréal, H3T 1V6   Canada

Choi, S. Chan,
Graduate School of Management,
Rutgers University,
180 University Av., Neward, NJ 07102-1895   USA

Cressman, George E.
DuPont - Marketing Development Group,
BMP 16-2108, P.O.Box 80016,
Wilmington, DE 19880-0016   USA

Dearden, Jim A.
Departement of Economics,
Lehigh University,
Bethlehem, PA 18015   USA

Desiraju, Ramarao
College of Business and Economics,
University of Delaware,
Newark, Delaware 19716-2710   USA

Divakar, Suresh
College of Business Administration,
SUNY at Buffalo, NY 14260   USA

Dockner, Engelbert
Institut für Betriebswirtschaftslehre,
University of Vienna,
Brünnerstrasse 72, A-1210 Vienna   Austria

Erickson, Gary M.
University of Washington,
School of Business,
Box 353200, Seattle, WA 98195-3200   USA

Gaunersdorfer, Andrea
Institut für Betriebswirtschaftslehre,
University of Vienna,
Brünnerstrasse 72, A-1210 Vienna   Austria

Hruschka, Harald
Faculty of Economics,
University of Regensburg,
D-93053 Regensburg   Germany

Jaffe, Morton I.
Baruch College,
The City University of New York
17 Lexington Avenue, New York 10010   USA

Jørgensen, Steffen,
Department of Management,
Odense University,
Campusvej 55, 5230 Odense M   Denmark

Koo Kim, Chung
Dept. of Marketing, Concordia University,
1455 de Maisonneuve, Montréal, H3G 1M8   Canada

Lilien, Gary L.
The Smeal College of BA,
The Pennsylvania State University,
University Park, PA 16802-3004   USA

Minowa, Yuko
P.O. Box 4766,
Highland Park, NJ 08904   USA

Monahan, George E.
Department of Business Administration,
University of Illinois,
Urbana-Champaign, IL 61280   USA

Napiorkowski, Gregory
AT&T Bell Laboratories,
600 Mountain Avenue,
Murray Hill, NJ 07974-0636   USA

Natter, Martin
Institut für Informationsverabeitung
University of Economics, Pappenheimgasse 35/3,
A-1200, Vienna   Austria

Sobel, Matthew J.
School for Management and Policy,
SUNY at Stony Brook,
Stony Brook, NY 11794-3775   USA

Thépot, Jacques
LARGE, Université Louis Pasteur,
38, Boulevard d'Anvers,
67070 Strasbourg Cedex   France

Yezza, Abdelwahab
GERAD, HEC
5255 Decelles,
Montréal, H3T 1V6   Canada

Yoon, Eunsang
Dept. of Marketing, College of Management,
University of Massachussetts at Lowell,
Lowell, MA 01854   USA

Zaccour, Georges
GERAD, HEC
5255 Decelles,
Montréal, H3T 1V6   Canada